PSYCHE AND SILICON: THE INTERPLAY OF ARTIFICIAL INTELLIGENCE AND HUMAN PSYCHOLOGY

MIND AND MACHINE: PSYCHOLOGICAL INSIGHTS INTO ARTIFICIAL INTELLIGENCE

NAKEL NIKIEMA

TABLE OF CONTENTS

INTRODUCTION

E ver heard of the principle of reciprocity?

It's a social construct that a psychologist named Dr. Robert Cialdini developed as one of his "seven universal principles of persuasion." According to his idea, matters operate effectively because of a two-way system.

For example, you're part of a small team at work, and you've been assigned a project that requires collaboration and input from everyone involved. To get the ball rolling, you volunteer to take on an extra task that falls outside your job description, knowing it will help the team succeed.

Because of your initiative and willingness to step up for them, your teammates are also more likely to step up for you—and reciprocate your actions.

So, whether you're aware of it or not, all of the many great things in life follow a give-and-take approach—a timeless concept that I'll use in this book to help you understand the interplay between the mind and machine.

PURPOSE OF THE BOOK

This book, *Psyche and Silicon: The Interplay of Artificial Intelligence and Human Psychology*, intends to let you peek closely at how AI and psychology work together.

I say this because it's apparent that AI is shaping the world. But, can it go as far as changing (or heavily influencing) how we think? How about how we feel? Or, how about the way we interact with others?

That's what you're about to find out! In this book, you can discover

- AI's history and evolution
- different types of AI
- AI's role in cognitive tasks
- how AI understands emotions
- affective computing
- human–computer interaction (HCI)
- psychological impact of AI on employment
- AI in mental health
- case studies of AI in healthcare, business, homes, and cities
- and more

This book is for everyone, and I'll accommodate different types of readers here. From psychologists who can learn new things about how AI is changing their field and AI researchers who can learn about psychological ideas that could inspire better artificial systems to students who can get a big-picture view of this new, somewhat radical, and growing area and people who make laws and can find all sorts of information to help create rules for AI.

It's for anyone interested in how AI affects society and wants to learn about these complex ideas in an easy-to-understand way.

I aim to help readers from all walks of life understand and brave this new AI-heavy world where AI and psychology are coming together. In writing this, I hope to make one thing certain: encourage people to think hard about how AI affects just about everything we do... our physical and mental health, how we make decisions, our behavior, how humans and machines interact, and our logic.

OVERVIEW OF ARTIFICIAL INTELLIGENCE

AI is moving beyond old-school algorithmic approaches to create more than capable systems. On their own, AI systems can surprise you by learning, adapting, solving complex problems, and more.

To say that things are moving at a fascinating speed may be an understatement. You can say that an AI system is practically a polymath. In addition to handling tasks (may they be mundane or extravagant).

At this point, the world has been introduced to AI concepts of sorts. We have machine learning (ML), deep learning, supervised learning, transfer learning, natural language processing (NLP), computer vision, robotics, and more.

What once were just theoretical concepts is now our reality. And, as AI is on track evolution-wise, there will be new related branches. I'm talking about explainable AI (XAI), edge computing, quantum ML, neuromorphic computing, and more. And, these are promising even greater capabilities and integration with human cognitive processes.

THE INTERSECTION OF AI AND PSYCHOLOGY

But, while AI has reached unparalleled heights, its evolution has only scratched the surface. With a market size poised to reach $407 billion in 2027, it has yet to hit its peak (Haan, 2024).

One of the evolving fields worth exploring is where AI and psychology meet. And, as mentioned earlier, this book will deliver just that.

As I go into certain topics, I'll also help you learn more about how human cognitive processes can be properly simulated, how interactions can be carried through with emotional intelligence, and how there is an underlying impact on human behavior and society.

Stitched between the fields of AI and psychology is an interdisciplinary approach. And, the good thing about it is it leads to a whole lot of more promising things.

With an interdisciplinary approach, you can anticipate innovation, an in-depth understanding of the most challenging problems, and a blend of solutions-oriented ideas. It can also encourage you to be (and stay) relevant in the real world and address different issues as they come.

Now, let's get to it!

FOUNDATIONS OF AI

Given that 75% (at least) of consumers are on the fence about the possibility of misinformation from AI in the early 2020s, an ideal way to move forward is to be educated.

— KATHERINE HAAN

For me, one of the most effective ways of learning something is to start from scratch. When you start from scratch, you can see things from the ground you're currently on, and then you go from there.

Starting from scratch is also (and quite arguably) an effective way to address misinformation. In a world where fake news and manipulated information can spread almost like wildfire, the ground-up approach lets you assess the validity of every claim and filter out misinformation.

My main objective is to assure you that the knowledge and understanding you develop is based on fact, not fiction. And, by building your understanding piece by piece, you're more likely to identify and address misinformation effectively.

My other objective is to set you up for the chapters ahead. Because we're going to talk in-depth about the role of psychology in all of this, I think it's best to clear up any misinformation about AI from the start.

The earlier, the better, right?

HISTORY AND EVOLUTION OF AI

According to research about the history and development of AI (conducted by folks from the University of Washington), you can trace AI's roots back to the 1950s. This was when pioneers like Alan Turing (the father of theoretical computer science) and John McCarthy (the father of AI) laid the theoretical groundwork (Smith et al., 2006).

Turing's seminal 1950 paper "Computing Machinery and Intelligence" proposed the Turing Test as a measure of machine intelligence. This spurred researchers to ponder how to create systems exhibiting humanlike cognitive abilities.

In 1956, McCarthy organized the Dartmouth Conference, widely thought of as the birthplace of AI. There, researchers like Marvin Minsky and Claude Shannon explored concepts like NLP, neural networks, and ML. Early AI systems focused on symbolic logic and rule-based reasoning to tackle problems head-on.

AI Winters and Boons: A Cycle of Hype and Sheer Disappointment

The history of AI has been marked by alternating periods of optimism and setbacks. The 1960s were a period of progress and bold predictions about how far AI can go.

But, it wasn't long until things went south. In areas like natural language understanding and computer vision, roadblocks came.

This brought in the first "AI winter" in the 1970s. It was when funding dried up in the middle of disappointment over a series of unfulfilled promises.

Sure, the expert systems boom of the 1980s brought renewed excitement, as rule-based systems showed promise in narrow domains. But again, limitations became apparent, and another AI winter set in during the 1990s.

Throughout these cycles, a lot of AI techniques continued to advance incrementally. Much to many people's dismay, though, progress was slow.

Modern AI

AI has experienced a dramatic resurgence since the early 2000s. Its transformation has been impressive—and nothing more.

Here are the factors that drove the resurgence:

- exponential growth in computing power (which inspired the development of more complex models)
- big data and improved algorithms for learning from massive datasets
- major breakthroughs in deep learning and neural network architectures

- cloud computing platforms that democratize access to AI capabilities

These all have led to remarkable achievements, especially in areas like computer vision, speech recognition, NLP, and game-playing. Modern AI systems can match or exceed human performance on many specific tasks.

One area that underwent staggering development in the modern-day scene is theorem-proving. A specific development is Isabelle. Isabelle is one of the most popular theorem provers and proof assistants. You can use it for formal verification, mathematical proofs, and computer science research.

But, we're still far from the kind of progress that the general intelligence envisioned by that of the early pioneers. There's no doubt that today's AI is excellent and has sharp skills in narrow, specialized tasks. Unfortunately, it's poised to get better in terms of the general reasoning abilities likened to human cognition.

And, as the capabilities of AI grow, so do concerns about how it's used. Many people show concern about its ethics, safety, moral standards, and societal impact. This is why the field now deals with almost unlimited challenges around bias, privacy, and the potential for job displacement.

As a simple resolution, modern AI systems are designed to be more than just your average run-of-the-mill systems that can provide you with basic assistance. Instead, they're equipped with features to show utmost fairness and accountability.

Regardless, the story of AI is one of rather extravagant and ambitious goals. It also underwent cycles of progress and setbacks, and gradual theoretical as well as technological advances. So, while we've come a long way from the heydays of AI research, in many ways the journey is just the start.

The good thing is that we have something worth looking forward to. The coming decades promise to bring even more innovations. As you—all of us, really—continue exploring and anticipating the heights of machine intelligence, there's a lot of hope for the future.

KEY CONCEPTS AND TECHNOLOGIES IN AI

Knowing the key concepts and technologies in AI provides you with the opportunity of a lifetime. With a good understanding of AI, you have the opportunity to develop a deeper psychological understanding of machine behavior.

Like I said in the previous chapter, matters tend to operate on a two-way system so, knowing how the technology operates is how you can gain complete awareness and a much better sense of how machines process and interpret data.

And, when all this is said and done, this helps you develop a kind of AI "intuition" that means you can work more effectively and collaboratively with these systems. You're excellent at predicting and anticipating their responses.

Here are the key concepts and technologies in AI that you should learn about.

ML

At the heart of modern AI is ML, which is an algorithm-based technique. It works by allowing systems to learn and improve through experience.

First, an AI system acquires data and then prepares it to let the algorithm process it for deployment. From there, its performance is assessed for necessary integration.

Here's a simple text-based diagram that shows the ML process in action:

[Data Acquisition]

↓

[Data Preparation]

↓

[Algorithm Choice]

↓

[Model Development]

↓

[Performance Assessment]

↓

[Production Integration]

↓

[Continuous Improvement]

Supervised Learning

Supervised learning is a common approach to ML. It involves using labeled datasets to train models that map inputs to known outputs.

This technique has numerous submethods, from the classic linear regression to more advanced methods like random forests, which involve combining multiple weak learners to create a more compelling ensemble model.

Unsupervised Learning

Another technique is unsupervised learning. Unsupervised learning works with unlabeled data. Algorithms in this category are designed to tease out patterns, structures, and relationships that may otherwise remain obscured.

Clustering techniques like k-means and density-based spatial clustering of applications with noise (DBSCAN) identify similar data points and group them, while dimensionality reduction methods such as t-stochastic neighbor embedding (t-SNE) and Uniform Manifold Approximation and Projection (UMAP) simplify complex data and extract important features.

Generative models like variable autoencoders (VAEs) and generative adversarial networks go even the extra mile. They work by learning to create entirely new data samples that mimic the distribution of the input data.

Reinforcement Learning

Reinforcement learning is another powerful technique. It is where an AI system interacts with its environment and learns to maximize a cumulative reward. This process is how AI can freely adapt and improve in complex, dynamic environments through a process of trial and error.

Reinforcement learning techniques range from classic Q-learning (where an agent learns the optimal value of each state in its environment) to advanced methods such as proximal policy optimization, which uses a policy gradient to improve the AI system's policy) and soft actor-critic, a reinforcement learning algorithm that combines policy gradients with value-based techniques).

Neural Networks

The architecture and functioning of artificial neural networks (ANNs) are describable when you have a good understanding of neural networks. These networks draw inspiration from biological neural systems, as they comprise interconnected nodes or "neurons" organized in layers.

The fundamental building block, the perceptron, aggregates weighted inputs and applies an activation function to produce an output. There are more advanced and modern ANNs, and they feature all sorts of diverse architectures that can handle specific tasks.

In neural networks, different architectures exist to suit various tasks. These range from convolutional networks optimized for image analysis to recurrent structures that can process sequential information.

Other examples include adversarial networks for data generation and transformer models that have revolutionized NLP. Each type offers unique capabilities, which presents researchers and practitioners with the perfect opportunity to select the most appropriate architecture for their specific application.

A feedforward network, the most basic structure, processes data in a single direction. If you think of it like a one-way street, information travels down the road with no U-turns allowed.

Convolutional Neural Networks and the Backpropagation Algorithm

For image processing, convolutional neural networks (CNNs) dominate. These networks have special layers and are called convolutional and pooling layers. Because of such layers, they can easily identify patterns, make connections, and extract features from images, even complex ones with multiple elements.

And, behind the bigger magic and streamlined efficiency of neural networks is the backpropagation algorithm, which serves as the engine of learning.

Backpropagation takes the loss function (a way to measure how well the network is performing) and uses it to figure out which weights and biases in the network need to be tweaked to improve performance.

NLP

As an important component of AI, NLP is another way for machines to understand human language. NLP techniques are versatile and span a wide range of tasks, from analyzing, simplifying, and interpreting text to generating easily readable and humanlike language.

A vital concept in NLP is tokenization. It revolves around breaking text into digestible, smaller units, or "tokens," that can be processed.

For example, in English, words are among the most common tokens. Other examples include numbers, symbols, and punctuation.

Robotics and Autonomous Systems

Robotics and autonomous systems are another aspect worth learning. They represent the physical embodiment of AI, and they integrate perception, decision-making, and action in real-world environments.

These systems operate on sensor fusion. They combine data from various sources to create a comprehensive environmental model.

Thanks to simultaneous localization and mapping (SLAM) algorithms, robots can navigate unknown spaces. And they do this by constructing maps while tracking their position.

Path planning algorithms, including sampling-based methods like rapidly exploring random trees (RRT) and optimization-based approaches like model predictive control, generate efficient and safe trajectories through complex, dynamic environments. Control systems, ranging from classical proportional–integral–derivative (PID) controllers to adaptive and robust control techniques, translate high-level commands into precise motor actions.

In autonomous vehicles, decision-making systems often employ a combination of rule-based approaches and ML models. Reinforcement learning techniques like Deep Q-Networks (DQN)and PPO let vehicles learn optimal driving policies.

Meanwhile, computer vision algorithms, often based on deep learning, facilitate object detection, semantic segmentation, and scene understanding.

TYPES OF AI: NARROW, GENERAL, AND SUPERINTELLIGENT AI

Different types of AI exist because there are various ways in which machines are trainable. Their types depend on how they can perform tasks or solve problems that would normally require human intelligence. This variety is the reason they can handle specific needs and scenarios.

Here are the different types of AI systems.

Narrow AI

Narrow AI, also known as weak AI, refers to AI systems engineered to perform specific tasks within a limited aspect. According to Investopedia, a global financial website, these systems excel at their designated functions (The Investopedia Team, 2022).

Unfortunately, they can't transfer knowledge or skills to other aspects. This lack of flexibility and generalization can limit the usefulness of narrow AI systems in solving problems or making decisions in new situations.

From a solution-focused perspective, this isn't a problem because these AI systems can still improve. Besides, research is ongoing in areas like transfer learning, which encourages people using AI systems to use knowledge learned from one task to improve their performance on related tasks (Hosna et al., 2022).

Here are the main characteristics of narrow AI:

- task-specific optimization
- reliance on predefined rules or ML models
- limited adaptability outside their trained field

- high efficiency within their specialization

General AI

General AI, or artificial general intelligence (AGI), is a hypothetical form of AI. It possesses humanlike cognitive abilities and can handle a wide range of tasks. Some people also think of it as the opposite of a weak AI system and is, therefore, considered strong AI.

AGI would be capable of solving new problems without human intervention. And, it'd be able to learn from experience, adapt to new situations, and even understand context and nuance in human communication.

While some people view AGI as the ultimate goal of AI research, others believe that this level of intelligence may never be achievable or is still far in the future.

Unlike narrow AI, AGI can

- reason abstractly
- problem-solve in new situations
- learn and adapt to new environments
- transferring knowledge between different fields

Superintelligent AI

Superintelligent AI, on the other hand, represents a theoretical stage of AI development where AI has gone above and beyond human cognitive abilities in all aspects. This concept goes beyond AGI or any strong system.

For regular or everyday users of AI systems, the concept of a superintelligent AI seems like a dream.

It also suggests intelligence that could

- rapidly self-improve its capabilities
- solve complex problems beyond human comprehension
- potentially redesign its architecture for optimal performance

If there is an AI system that is perceived to be a threat, it's a super-intelligent AI. Beyond this, the other staggering concern is that there are significant technical limitations to this form of AI. At this point, it's still hypothetical.

One significant limitation is the complexity and cost of developing and maintaining superintelligent AI systems. These systems can perform extremely well, but for their performance to be possible, they would require vast amounts of computing power and resources. And, this just means that they're challenging and expensive to develop and operate.

Plus, the algorithms and methods required to develop superintelligent AI may not yet exist. This says that there is a need for significant research and development before this could happen.

COGNITIVE PSYCHOLOGY AND AI

You've probably heard of the concept of working memory capacity, right? Well, it's a cornerstone of cognitive load theory and has implications for the design and administration of IQ tests.

According to research that follows up on this theory's development, test creators must take initiative and avoid overwhelming the test-taker's working memory. That means absolutely no overly complex or demanding tasks (Moreno & Park, 2012).

The reason:

If a test item requires excessive cognitive load, it may result in a lower test score. And, mind you, this isn't because of a lack of intelligence. Rather, this is because of the test-taker's limited working memory capacity.

Now, this understanding shows the importance of careful test design in accurately measuring IQ. Importantly, it also shows how AI systems perform.

Just as human test-takers can struggle with tasks that exceed their working memory capacity, AI systems can struggle with tasks that overwhelm their memory or computational resources.

As AI systems become increasingly complex, designers must carefully consider the cognitive load of their tasks. They need to figure out a way for the system to perform as intended without being hampered by overburdened memory or any sort of processing limitations.

After you've learned about the foundations of AI in the previous chapter, let's now move forward and discuss everything you need to know about cognitive psychology and AI.

COGNITIVE MODELS AND AI

Ignoring the significant strides AI has made in replicating human cognitive processes is almost impossible. AI systems are becoming more advanced as time goes on.

When it comes to their ability to mimic various aspects of human thought, it's becoming a challenge to differentiate which is AI and which is the real deal.

Simulating Human Cognition: How Do AI Models Replicate Human Cognition?

According to Pearce (2023), AI models—diffusion models in particular—can replicate human behavior in interactive environments. They can do so by learning from datasets of human observations and actions.

These models can capture the full diversity of human behaviors. They can also pave the way for the generation of varied and humanlike responses in practically every task there is!

Here are the areas where AI models can replicate human behavior:

Area	The Process
Attention and focus	Attention mechanisms in AI, such as those used in Generative Pre-trained Transformer (GPT) models, replicate human selective attention.
	These models can focus on relevant parts of input data and completely ignore irrelevant information.
Decision-making	Reinforcement learning algorithms replicate human decision-making processes.
	For instance, in game-playing AI, these algorithms learn optimal strategies by balancing exploration (trying new moves) and exploitation (using known successful strategies), mirroring how humans learn through trial and error.
Emotional recognition	While not replicating emotions themselves, AI models can simulate the process of recognizing emotions.
	Multimodal models combine visual and auditory inputs to detect emotional states from facial expressions, voice tone, and body language.
Language understanding	Transformer models, such as Bidirectional Encoder Representations from Transformers (BERT), simulate human language processing by considering the context in both directions.
	They use self-attention mechanisms to weigh the importance of different words in a sentence, similar to how humans focus on the main elements when interpreting language.
Memory and learning	Long short-term memory (LSTM) networks work like a human being's working memory.
	They can maintain relevant information over time while discarding less important or totally irrelevant data.
Problem-solving	AI systems like AlphaFold simulate human problem-solving in complex domains.
	By combining deep learning with domain-specific knowledge (in this case, protein folding), these models can tackle problems that require both pattern recognition and logical reasoning.

Visual processing	CNNs mimic the human visual cortex.
	They use hierarchical layers to process visual information, starting with basic features like edges and colors, and then progressing to more complex patterns.

While these AI models have made strides in replicating human behavior and specific cognitive processes, they still lack the generalizable intelligence and consciousness characteristic of human cognition. In a way, it's unfortunate.

The good news is that current research is here to save the day (Celemin et al., 2022). After all, it focuses on integrating these individual capabilities into more comprehensive models. Above many things, it's poised to create AI systems that can flexibly apply cognitive skills and bring us closer to AGI.

Information Processing Theory: Human vs. Machine

Information processing theory is a framework for understanding humans and machines. It's about knowing the reason behind the way they handle, store, and utilize information. While there are similarities between human and machine information processing, there are also significant differences in their approaches and capabilities.

Human Processing

In human information processing, sensory input is first received through various channels (sight, sound, touch, and more), and from then on, it's briefly stored in sensory memory. Attention mechanisms then filter this information and figure out what moves to leave to short-term (or working) memory.

This stage is limited in capacity and duration. During this period, information considered important is then encoded into long-term memory through rehearsal, elaboration, and other similar processes.

Machine Processing

Meanwhile, machine information processing, particularly in AI systems, follows a somewhat analogous path. Input data is received through sensors or data feeds, and then processed through layers of a neural network or other computational structures. The system's "attention" is directed by programmed algorithms or learned patterns.

Short-term processing occurs in random-access memory (RAM). But, when it comes to long-term storage, various forms of computer memory or databases are used.

ML AND HUMAN LEARNING

Human learning and ML share fundamental principles. However, they diverge significantly in their implementation and capabilities. Both systems may be targeted to improve performance through experience. There are underlying mechanisms, though, and these mechanisms differ substantially.

Learning Mechanisms: Characteristics, Similarities, and Differences

In human learning, the brain's neural networks form and strengthen connections through synaptic plasticity. This process involves complex biochemical changes. It's the reason for the encoding of information and skills.

Characteristics

It follows that humans learn through various methods, which commonly include observation, trial and error, and explicit instruction. On the flip side, ML is about algorithms and statistical models to identify patterns in data.

Similarities

A main similarity between human learning and ML is the importance of feedback. Humans turn to various forms of feedback to refine their understanding and skills.

Meanwhile, ML algorithms use error signals or reward functions. This is what they use as the basis to adjust their parameters and improve performance.

Differences

A significant difference is in the speed and scale of learning. ML systems can, indeed, process vast amounts of data in almost a fraction of a second. This is why they can excel in specific, well-defined tasks.

Humans, while slower in data processing, exhibit astounding efficiency in learning from a few examples and transferring knowledge. Some people have exceptionally high intelligence and can effortlessly learn something fast.

Another distinction is representation. Human knowledge is often abstract, flexible, and context-dependent. For this reason, human logic is used for creative problem-solving and analogical reasoning.

ML models operate on more rigid, mathematical representations. While this makes them highly effective for specific tasks, they may struggle with nuanced, context-dependent information.

The interpretability of learning also differs. Human learning processes are often intuitive and difficult to fully explain, while ML models can be analyzed and debugged more systematically, although some advanced models (like deep neural networks) present challenges in terms of interpretability.

Transfer Learning and Generalization: The Power of AI and Humans in New Situations

Transfer learning and generalization are almost indispensable abilities for both human and artificial intelligence. This lets them take the application of previously acquired knowledge to new heights. And, while humans excel and are "naturals" at these tasks, AI systems aren't ones to be underestimated because they're catching up.

The Power of Humans

In human cognition, transfer learning occurs effortlessly. We can apply abstract principles learned in one area to solve problems in entirely different fields. This ability is rooted in our capacity to form mental models and recognize connections and underlying patterns that may be easily missable or transcend specific contexts.

For instance, a person who knows the nitty-gritty about the concept of supply and demand is a person who can prevent themselves from getting stuck in traffic. Particularly, they can apply this principle to get a sense of traffic flow in urban areas.

The Power of AI

In a traditional sense, AI systems have been more limited in their ability to transfer knowledge. Classic ML models are often task-specific. This causes them to require retraining for each new application. Then again, the more recent advances in transfer learning techniques have dramatically improved AI's "superpower" in this area.

One popular approach in AI transfer learning is fine-tuning pretrained models. This method is how the model can use and take advantage of the general language understanding for specialized applications. In turn, this significantly reduces the amount of task-specific data required.

Another technique is domain adaptation. This is where models are trained to perform well on a target domain using knowledge from a related source domain. This comes in handy when labeled data in the target domain is scarce.

AI IN COGNITIVE TASKS: PERCEPTION, MEMORY, AND DECISION-MAKING

Unlike humans, AI systems have unwavering attention. They can process never-ending amounts of data without fatigue or distraction. And, this causes them to be adept at identifying patterns and making decisions with a high degree of accuracy.

They also have dedicated processing power. The computational resources available to AI systems enable them to perform tasks that would be impossible for humans to do manually, such as analyzing massive amounts of data or conducting simulations in real time.

Below, you'll find relevant discussions that point to the reason AI excels in cognitive tasks.

Perception

AI-driven perceptual systems have become among the next big things in terms of machines and their understanding of visual and auditory inputs. In computer vision, for one, advanced deep learning architectures are the essential structures in image processing and analysis.

These advanced models use multiple layers of artificial neurons. Their objective is to extract and interpret visual features, which makes them work more effectively from the most basic elements to highly complex object recognition.

For auditory perception, AI uses techniques like spectral analysis and recurrent networks to process time-dependent audio signals. It also uses deep learning models, particularly transformers, to make the most out of improved speech recognition and language processing.

Memory

AI memory models have evolved significantly and are now better than ever. They're no longer just systems for simple data storage and are rather a close call to human memory functions.

In traditional ML, information is typically stored in model para-meters or lookup tables. But, with these advanced AI systems, you can find more dynamic and efficient memory structures.

If you're familiar with memory networks and neural Turing machines, you can understand (and be impressed by) how these systems can introduce external memory components that are readable from and written to.

Because of this, more flexible information storage and retrieval is no longer taboo. Best of all, these models can perform complex reasoning tasks and all they need to do is to manipulate their memory contents.

Differentiable neural computers (DNCs) take this further. They go to the next level by incorporating a memory matrix with read-

and-write operations. This means that they can solve complex, structured tasks.

These advanced memory models are the secret as to why AI systems can seem to do it all. Thanks to their design, AI can handle extravagant tasks or tasks that require long-term context, sequential reasoning, and quick adaptation to new information.

Decision-Making

Throughout the years, AI decision-making processes have advanced. Long gone were the days when rule-based systems reigned supreme. Because of advancements after advancements in AI, strategic decision-making is now possible.

For strategic decision-making, Monte Carlo tree search (MCTS) has proven highly effective. For example, AlphaGo's victory over human Go champions. This algorithm balances the exploration of new strategies with the exploitation of known good moves.

In uncertain environments, Bayesian decision theory provides a framework for AI to reason about probabilities and utilities. Probabilistic graphical models, such as Bayesian networks, lead to the structured representation of complex decision problems.

Meanwhile, for multi-agent scenarios, game theory concepts are integrated into AI decision-making processes. The reason for this is to allow strategic reasoning in competitive or cooperative settings.

In real-world applications, on the other hand, ensemble methods combine multiple models or decision algorithms to improve robustness and performance.

EMOTIONAL AI

The *Emotion Machine* is this mind-bending book that came out back in 2006, authored by Marvin Minsky, one of the AI pioneers (mentioned in Chapter 2) and the person who cofounded MIT's Artificial Intelligence Laboratory.

In the book, Minsky explores how our emotions work and how we might actually recreate them in machines.

He proposes that we can break down emotions into mathematical formulas. And, if we crack that code, well, we could create extraordinary machines... machines that don't just process information but actually feel in a meaningful way.

Now, here's where it gets even more intriguing: Minsky doesn't stop at the "how." He also explores the "what if."

What happens when we have machines that can understand and express emotions?

If we can create AI systems that understand and process emotions, we're not just talking about smarter machines. We're talking about machines that can interact with us on a whole new level.

So, as we move on from our exploration of cognitive psychology and AI, remember the book and what it stands for. Our discussions in this chapter should help you get a better understanding of emotional AI and what you can do to use the power of emotional intelligence to your advantage.

UNDERSTANDING EMOTIONS IN HUMANS AND MACHINES

AI systems work in a pretty straightforward manner. While it may seem like they're just churning out results like clockwork, a lot is going on behind the scenes.

Theories of Emotion

By using emotional understanding in your AI development, you can make the system work better for you. When you handle things correctly, it responds more effectively to your requests.

That said, it's important to have a good grasp of the different theories of emotions. It's one of the most effective ways you can gain a good understanding of how AI systems interpret processes.

Here's a list of the theories that you need to be aware of:

James–Lange theory

- **Overview:** This theory says that our emotions come from how our body reacts to things around us.
- **Example:** We feel happy because we are smiling, or we feel fear because we are trembling.

Cannon–Bard theory

- **Overview:** This theory posits that emotions and physiological reactions occur simultaneously and independently when we encounter a stimulus.
- **Example:** We feel afraid at the same time our heart races when we see a snake.

Two-factor theory (Schachter–Singer theory)

- **Overview:** This theory states that emotions are based on physiological arousal and the cognitive interpretation of that arousal in context.
- **Example:** If our heart races in a dark alley, we may label that arousal as fear; if it races at a concert, we might interpret it as excitement.

Cognitive appraisal theory

- **Overview:** According to this theory, our emotions arise from our interpretation and evaluation of situations, which determines their emotional significance to us.
- **Example:** We feel happy about receiving a promotion after perceiving it as a reward for hard work, while we might feel anxious about a job loss due to the perceived threat to our stability.

Evolutionary theory of emotion

- **Overview:** This theory suggests that emotions have evolved to serve adaptive functions, helping us respond to challenges and enhancing survival.

- **Example:** Fear triggers a fight-or-flight response to danger, while joy fosters social bonds and cooperation, which can benefit group survival.

Emotion Recognition

Understanding emotions is key to building and maintaining relationships. In personal interactions, recognizing emotions can help you respond appropriately.

For example, if a friend is upset, acknowledging their feelings can show empathy and support. This creates a bond and fosters mutual trust. In the workplace, being able to read your colleagues' emotions can also improve communication.

Traditional Techniques

The old-school approaches still work. Sure, it can be easy to overlook the wisdom of traditional methods in favor of newer, flashier techniques. But, many old-school approaches have stood the test of time.

These methods often give you a foundation that can enhance modern practices. Understanding how they work can bring you insights and benefits.

Here are traditional emotion recognition techniques:

Technique	Description
Facial expression analysis	Identifying emotions based on facial movements and expressions.
Voice analysis	Detecting emotions through vocal features like tone and pitch.
Body language interpretation	Analyzing posture and gestures to infer emotional states.
Text analysis	Using NLP to determine emotions in written text.

Physiological measurement	Monitoring biometric indicators like heart rate and skin conductance.
Eye tracking	Analyzing eye movement and gaze direction to understand emotions.
Behavioral pattern Recognition	Observing patterns in actions over time to identify emotions.
Sentiment analysis	Evaluating the sentiment behind words and phrases.
ML models	Using AI to classify emotions based on input data.
Multimodal approaches	Combining multiple methods to achieve accurate emotion recognition.

Modern Techniques

Modern problems call for modern solutions. While traditional approaches work, sometimes a different way is more effective.

I'm saying this because we live in a modern world right now. And, right now, things are different and to exist in an environment with them around, it means that we have to take new and different considerations.

Here is a table that features a list of modern emotion recognition techniques:

Technique	Description
Neurofeedback	Using brain activity feedback to help regulate emotional responses.
Social media behavior analysis	Analyzing online interactions to gauge emotional states.
Wearable technology	Monitoring physiological changes to detect emotional shifts.
Virtual reality (VR) simulation	Engaging individuals in VR to analyze emotional responses.
Digital footprint analysis	Tracking online behaviors to infer emotional states.
Music preference assessment	Analyzing music preferences to understand mood and emotions.
Storytelling analysis	Evaluating storytelling to uncover underlying emotional states.
Art and creativity analysis	Understanding emotions through creative expressions and artworks.

Robot interaction analysis	Observing interactions with robots to analyze emotional responses.
Dream analysis	Analyzing dreams to understand subconscious emotional states.

AFFECTIVE COMPUTING: AI AND EMOTIONAL INTELLIGENCE

Affective computing and emotional intelligence are two terms that you might hear often. They're closely linked, yet they fulfill distinct roles in how we understand and engage with emotions. To unpack these ideas, it helps to look at each term separately while recognizing the overlap in their applications.

Affective Computing and Emotional Intelligence: The Connection Between the Two

Affective computing is about creating systems that can understand and respond to human feelings. This study aims to build technology that can react to emotions.

Meanwhile, emotional intelligence is the ability to see, understand, and control your emotions and the emotions of others. It is important for building strong personal and work relationships. It includes skills like empathy, which is understanding how others feel, self-regulation, which is managing your own emotions, and social awareness, which is being aware of what's happening in social situations.

Here's a discussion of their connection.

Empathy Simulation

Through emotion simulation, affective computing systems can mimic feelings, allowing them to act more like humans. This helps users see the system as emotionally smart.

Adaptive Interactions

Affective computing apps can change how they respond based on the emotions they sense, leading to better and more understanding interactions.

For example, a virtual tutor can notice when a student is having a hard time and change how it teaches. This ability helps the tutor give personalized help, keeping the student interested and motivated.

Improved User Relations

By understanding and responding to emotions, affective computing tools can foster deeper and more meaningful interactions, similar to those seen in emotionally intelligent human relationships.

Training and Development

Affective computing technologies can help people learn emotional intelligence. These tools give feedback on how to recognize and manage emotions, so users can improve their emotional skills.

Emotional AI Systems

By understanding users' emotional states, these systems can tailor responses and recommendations, creating a more personalized and engaging experience. This is beneficial in areas such as ecommerce, education, and entertainment.

Here are common emotional AI systems:

1. Affectiva

- **System:** Affectiva's emotion recognition technology analyzes facial expressions and vocal tones to gauge emotional responses.
- **Use:** This technology is used in various applications, including market research, automotive safety, and mental health monitoring.

2. CereProc

- **System:** CereProc develops emotional text-to-speech software that conveys different emotional tones through synthetic voices.
- **Use:** The technology has applications in gaming, virtual assistants, and accessibility tools, making digital interactions more relatable.

3. Realeyes

- **System:** Realeyes uses computer vision and ML to track facial expressions and measure emotional engagement in response to visual content.
- **Use:** This system is commonly utilized in advertising and marketing to analyze audience reactions and optimize campaigns.

4. Emotient

- **System:** Emotient, now part of Apple, developed technology to recognize emotions from facial expressions in real time.

- **Use:** The technology was used in healthcare, retail, and other industries to enhance customer experiences and monitor emotions.

5. Beyond Verbal

- **System:** Beyond Verbal analyzes vocal emotions, interpreting the emotional state of a person based on their voice.
- **Use:** This technology has applications in health monitoring, customer service, and enhancing virtual assistants.

Applications of Emotional AI: Healthcare, Customer Service, and Education

With emotional AI, an increasing number of companies are using it to enhance their services and engage more deeply with their customers. Companies are specifically using emotional AI to drum up the value they provide by personalizing interactions, understanding customer feelings, and improving overall customer experience.

This technology also allows businesses to better understand and respect diverse cultural expressions. This encourages them to raise awareness of cultural nuances, promote traditional practices, and create products that resonate with local communities.

Emotional AI in Healthcare

Emotional AI has the potential to transform patient care in healthcare settings. Studies show that 70% of healthcare professionals believe that emotional AI can lead to better patient engagement and satisfaction (Jennings & Knapp, 2023).

Here are note-worthy applications:

- **Mental health monitoring:** Emotional AI can analyze patient interactions, whether through speech or facial expressions, to detect signs of anxiety, depression, or stress. This information can provide mental health professionals with valuable insights for better diagnosis and treatment planning.
- **Personalized Patient Experience:** By understanding a patient's emotional state, healthcare providers can tailor their approach and communication style to fit the needs and preferences of the individual. This personalized care can improve patient satisfaction and adherence to treatment plans.
- **Telehealth and virtual therapy:** In online meetings, emotional AI helps recognize a patient's feelings during video calls. This helps therapists and doctors give the right emotional support and feedback, making remote therapy sessions more effective.
- **Patient behavior analysis:** Emotional AI can analyze patterns in patient emotions through wearable devices or mobile apps, offering insights into mood changes and overall well-being. This information can help healthcare providers identify at-risk patients and intervene early.
- **Medication adherence:** By monitoring emotional responses and engagement levels, emotional AI can help identify patients who may struggle with medication adherence. The technology can trigger reminders or motivational messages tailored to the emotional context of the patient.
- **Enhancing patient–provider communication:** Emotional AI tools can provide insights into a patient's emotional

state, enabling healthcare providers to adjust their communication strategies. This can foster better rapport, trust, and understanding between patients and providers.

- **Training and support for healthcare professionals:** Emotional AI systems can be used to train healthcare providers in recognizing and responding to patients' emotional needs. Simulated patient interactions can help develop empathy and emotional intelligence skills, which are crucial for effective patient care.
- **Predictive analytics for health outcomes:** By combining emotional data with other health information, healthcare systems can create models to predict health outcomes like possible hospital readmissions or complications. This helps in taking action before problems arise.

Emotional AI in Customer Service

Emotional AI is changing customer service by helping businesses understand and react to how customers feel. By creating more personal and caring interactions, companies can improve customer happiness, and loyalty, and keep them coming back.

Here are some notable applications:

- **Sentiment analysis:** Emotional AI looks at customer messages—like emails, chat messages, and posts on social media—to understand their feelings. This helps customer service workers see how customers feel about their experiences, which leads to kinder and more understanding responses.
- **Personalized customer experience:** By understanding a customer's emotional state, companies can shape interactions to better meet individual needs.

- For example, if a customer is frustrated, a representative can adopt a more empathetic and calming approach to resolve their issue effectively.
- **Emotion-driven routing:** Emotional AI can help route customer inquiries to the most appropriate representatives based on the detected emotion.
- For instance, a call from an upset customer can be prioritized and directed to a specialist who is skilled at handling stressful situations.
- **Virtual assistants and chatbots:** Emotional AI technology can be integrated into virtual assistants and chatbots, allowing them to recognize and respond to customer emotions. This leads to more engaging and humanlike interactions, which can improve customer satisfaction during automated support.
- **Customer feedback analysis:** By analyzing feedback surveys and reviews, emotional AI can extract insights about customer emotions related to products or services. This feedback can inform improvements and drive strategies to enhance customer experiences.
- **Proactive customer support:** Emotional AI systems can monitor customer interactions and detect signs of frustration or dissatisfaction. This enables businesses to act proactively by reaching out to affected customers before they escalate their issues.
- **Enhanced training for support staff:** Emotional AI can be used to analyze customer interactions and coach support staff on recognizing and managing emotions effectively. Training simulations can help develop soft skills like empathy and conflict resolution, improving overall service quality.
- **Predictive analytics for customer retention:** By analyzing emotional signals, companies can identify

customers at risk of churn based on negative emotional experiences. This allows businesses to implement targeted strategies to retain those customers.

- **Understanding customer preferences:** Emotional AI can help identify underlying emotional motivations behind customer preferences and behaviors. This understanding can drive personalized marketing strategies and improve product recommendations based on individual emotional responses.

Emotional AI in Education

Emotional AI offers numerous applications in education, from personalizing learning experiences to enhancing teacher–student relationships. By leveraging emotional insights, educators can create supportive and engaging environments that foster student success and well-being.

Here are some notable use cases:

- **Emotion recognition in the classroom:** Emotional AI can analyze students' facial expressions and body language to gauge their emotional states during lessons. This insight allows teachers to identify when students are confused, bored, or engaged, enabling them to adjust their teaching methods accordingly.
- **Personalized learning experience:** By understanding individual students' emotional responses and engagement levels, educational platforms can adapt content, pacing, and instructional strategies to meet the specific needs and preferences of each student.
- **Real-time feedback for educators:** Emotional AI technologies can provide teachers with real-time feedback about class dynamics and student emotions, helping them

to manage classroom environments more effectively and respond to students' emotional needs promptly.

- **Virtual learning environments:** In online learning settings, emotional AI can monitor student engagement and emotional states through video interactions. This data can help educators identify when students may be struggling and offer support or resources to enhance learning outcomes.

- **SEL tools:** Educational platforms can integrate emotional AI to facilitate social-emotional learning (SEL) by helping students recognize and manage their emotions. These tools can provide exercises, feedback, and strategies for developing emotional intelligence.

- **Gamified learning experiences:** Emotional AI can enhance gamified educational tools by adapting challenges and rewards based on the player's emotional engagement. This increases motivation and keeps students invested in their learning journey.

- **Early intervention for at-risk students:** By analyzing emotional signals and engagement patterns, educators can identify students at risk of academic failure or mental health issues. This enables timely intervention and support to help these students succeed.

- **Improving communication with parents:** Emotional AI can help schools analyze feedback from parents, capturing emotional tones in communications. This understanding can inform better strategies for engaging parents in their children's education positively.

- **Teacher training and development:** Emotional AI can help teachers learn emotional skills and understand how important feelings are for managing classrooms and interacting with students.

HUMAN–COMPUTER INTERACTION

~~~

Creativity is just connecting things.

— STEVE JOBS

T he HCI field focuses on how people use technology. It aims to create easy-to-use interfaces that help users interact well with machines, allowing them to express their creativity.

By understanding how users think and behave, designers of HCI strive to develop tools that enhance creativity, allowing individuals to draw connections between their thoughts and the digital world. In this context, the act of creativity can be seen as an interaction where innovative ideas are sparked not only through personal insight but also through the effective use of technology, leading to new ways of connecting and creating.

So, after going over emotional AI systems in the previous chapter, let's now talk about HCI. This way, you can understand how these systems can enhance user experience (UX) and facilitate more intuitive and effective communication between humans and machines.

PSYCHOLOGICAL PRINCIPLES OF HCI

HCI is a fascinating field that focuses on how people engage with computers and other digital devices. This area of study seeks to improve the relationship between users and technology, making it more efficient and enjoyable.

The main goal of HCI is to understand how people behave, what they feel, and what they need when using different digital tools. This information is very helpful for designers, allowing them to create products that work well and are easy for people to use.

The Role of Psychological Principles in HCI

Psychological principles include understanding cognitive processes such as perception, memory, and problem-solving.

For instance, when designing an interface, it is important to consider how users perceive information. Visual elements, like color and layout, can greatly affect how quickly and easily users can understand what they see on the screen. Designers often rely on established psychological theories to guide their decisions.

One example is the Gestalt principles, which describe how people naturally group visual elements together. By applying these principles, designers can create interfaces that feel intuitive to users.

Creating Better Interfaces and Reducing Cognitive Load

To create better interfaces, designers must prioritize the UX. This involves researching to gather insights about the target audience.

Once designers understand their users better, they can design the interface to meet their specific needs.

For instance, if data shows that users struggle with finding certain features, designers might simplify the navigation or make those features more prominent. By directly addressing user challenges, the design becomes more effective and satisfying.

Here are ways you can reduce cognitive load:

Method	Description
Chunking	Organize information into smaller, manageable units or "chunks" to make it easier to process and remember.
Simplification	Remove unnecessary elements or distractions from the interface, focusing on what is essential for the task.
Consistency	Maintain consistent layouts, terminology, and design elements throughout an application to help users learn quickly.
Use of visuals	Incorporate diagrams, images, and infographics to convey information more efficiently than text alone.
Progressive disclosure	Present information in layers and reveal only the necessary details for each step, reducing overwhelm.
Clear navigation	Design intuitive navigation menus that allow users to easily find what they need without excessive searching.
Feedback and affordances	Provide clear, immediate feedback for user actions to help them understand the results of their interactions.
Pretraining and familiarization	Offer tutorials or introductory materials to prepare users before they engage with a complex system.
Limit choices	Reduce the number of options presented at any given time to avoid decision fatigue and simplify the decision-making process.

Improving UX

Improving UX is at the core of HCI. A positive UX can lead to greater satisfaction, increased productivity, and returning customers. One way to improve UX is through user-centered design.

This method puts the user first in the design process, ensuring their needs and preferences are prioritized. For one, designers often create fictional characters known as personas, which represent various types of users.

For example, one persona might be "Sarah," a busy working mother who values quick and efficient access to information. By understanding Sarah's behaviors and preferences, designers can make strong and well-rounded decisions that cater to her needs, such as simplifying navigation or enhancing mobile usability.

These personas serve as a guiding framework, helping to maintain attention on what users truly want throughout the development phase.

Examples of HCI in Everyday Life

HCI can be seen in many aspects of everyday life. For instance, think about how we use smartphones. Most people rely on their phones for numerous tasks, from texting and browsing the internet to managing their calendars. The design of smartphone interfaces is important for ensuring ease of use.

According to a survey, 73% of people like websites that work well on mobile devices (Kowalski, 2021). This shows how important it is for websites to be easy to use.

Features like touchscreens and voice commands allow users to interact with their devices naturally. Companies like Apple and Google invest heavily in HCI research to ensure their devices are intuitive and user-friendly.

Importance of Inclusivity in Design

Inclusivity is another important consideration in HCI. Designers should strive to create interfaces that accommodate diverse users, including those with disabilities.

This could involve implementing features like screen readers for visually impaired users or ensuring that the interface is navigable by keyboard for those who cannot use a mouse.

Accessibility guidelines are rules created to help make websites, apps, and online content easy for people with different disabilities to use.

The most widely known rules are the Web Content Accessibility Guidelines (WCAG), created by the World Wide Web Consortium (W3C).

Here are some key aspects of accessibility guidelines:

Principle	Description
Perceivable	Information must be presented in a way that users can perceive, including text alternatives for non-text content and adaptable layouts.
Operable	User interface (UI) components must be navigable by all users, ensuring keyboard accessibility and providing enough time to interact with content.
Understandable	Content and UI operations must be clear and straightforward, using simple language and predictable behaviors.
Robust	Content must be compatible with current and future user agents, including assistive technologies, by using standard HTML and CSS.
Level A, AA, AAA	WCAG guidelines are categorized into three conformance levels, with Level AA being the recommended standard for most websites.

USABILITY AND UX IN AI SYSTEMS

Testing the usability and UX in AI systems is important. When we talk about usability, we mean how easy and efficient it is for users to interact with an AI system. A system that is not user-friendly can lead to frustration, mistakes, and a lack of trust.

On the other hand, UX encompasses the overall feelings and perceptions a user has when interacting with an AI. This includes everything from how intuitive the interface is to how well the AI understands user needs and preferences.

The Importance of Usability Testing

In AI systems, usability testing can help identify issues that users might face while interacting with the software. This allows them to gather valuable feedback on how the AI performs in real-world situations.

For example, if a user has to go through multiple steps to get a simple answer from an AI chatbot, this can be a major usability issue. To conduct usability testing, companies often use methods like user interviews or observing users as they interact with the system.

A good strategy for usability testing is to involve diverse users in the process. Different people may have varying levels of experience with technology, which can affect how they interact with AI.

By including a range of users in the testing phase, developers can gain insights into common issues and areas for improvement. After observing users, they can categorize the problems they face and prioritize which issues to address first.

Methods for Testing

You can use several methods to assess both usability and UX in AI systems, which include surveys, usability testing, and analytics, which have been shown to improve user satisfaction.

Here is a compilation of the methods you can use:

- **A/B testing:** Implement A/B testing to compare different versions of the AI system or features, determining which version performs better in real-world scenarios.
- **Benchmarking:** Compare the AI system to previous models to see if it performs better.
- **Compliance audits:** Ensure that the AI system meets regulatory and ethical standards, conducting audits to verify compliance with policies and guidelines.
- **Cost-benefit analysis:** Assess the economic impact of implementing the AI system by comparing the costs of deployment and maintenance with the benefits it provides.
- **Error analysis:** Analyze the errors made by the AI system to understand its limitations and areas for improvement.
- **Explainability and interpretability:** Assess how understandable the AI system's decision-making process is to users and stakeholders, ensuring it can provide clear explanations for its outputs.
- **Longitudinal studies:** Perform studies over an extended period to evaluate how the AI system performs in changing environments and its long-term effectiveness.
- **Performance metrics:** Use quantitative metrics such as accuracy, precision, recall, F1 score, and area under a receiver operating characteristic (AUC-ROC) to evaluate how well the AI model performs its tasks.
- **Robustness testing:** Evaluate the AI system's robustness by testing it against adversarial examples or unexpected input to see how it handles edge cases.
- **User studies:** Conduct user testing sessions to gather qualitative feedback on how users interact with the AI system and assess its usability.

Heuristic Evaluations

Usability heuristics are helpful in interaction design. They are general guidelines that assess UIs.

These rules aim for one thing: Make the UX better. They do this by ensuring designs are easy to use, quick, and effective. By following these rules, designers can create UIs that clearly meet users' needs.

What Are Usability Heuristics?

Usability heuristics are guidelines meant to make products easier to use. They are suggestions, not strict rules, that point out important parts of how users interact with designs.

One common guideline is the visibility of system status. This means that users should always know what is happening on the screen.

For example, when a user saves a document, a clear message like "Saving..." should show up. This message lets users know their action is being processed, which helps ease their worries about whether the system is working.

The Role of Consistency

Another important usability heuristic is consistency. Consistency refers to the idea that similar actions should produce similar results, which minimizes confusion.

When a user interacts with various elements of a design, they expect a sense of familiarity. This means that they want similar buttons or features to behave the same way across different parts of an application.

For example, if a "Submit" button is green in one area of the site, it should also be green everywhere else. This consistency helps build user familiarity with the interface, making it easier to navigate without needing to constantly relearn how the design works.

Error Prevention

Error prevention is another critical aspect of usability heuristics. This principle emphasizes designing interfaces that help avoid mistakes before they happen.

For instance, when users fill out a form online, the system can help prevent errors by checking the format of the input in real time. If an email field requires a specific structure, like having an "@" symbol, the system can provide immediate feedback when a user enters an incorrect format.

User Control and Freedom

Usability heuristics also advocate for user control and freedom. This principle suggests that users should feel in control of their interactions with the interface. Providing the option to undo actions or navigate back to a previous state allows users to feel more secure in their choices.

For instance, if a user accidentally deletes a file, having an "Undo" button can prevent them from panicking. Giving users this freedom encourages exploration and experimentation within the design.

Affordances and Signifiers

Affordances and signifiers play an essential role in usability heuristics. An affordance suggests how an object can be used, while a signifier indicates where the action should take place.

For example, a button that looks raised suggests it can be pressed. When users see a button that appears clickable, they can intuitively understand that it will act. Properly designing these elements ensures users can easily understand how to interact with the interface.

The Importance of Feedback

Feedback is an important element of usability, not just in terms of system status but also in the overall UX. Research indicates that effective feedback (as part of great design) can increase user satisfaction by up to a whopping 200% (Solomons, 2023). Users appreciate knowing that their input is valued and that their actions yield results.

Designing interfaces where users can give feedback easily, such as through surveys or ratings, invites a dialogue that can guide future improvements. Showing users that their opinions matter encourages a sense of community and ownership over the product.

Iterative Design Process

Usability rules should be part of the design process. This means that designs should be improved regularly based on what users say and how they use them.

By adopting an Agile approach, designers can make small, incremental improvements that lead to a better overall product. This flexibility allows teams to pivot as needed and respond to user needs more effectively over time.

AI AND SOCIAL PSYCHOLOGY

> *As AI systems become more sophisticated in modeling human behavior, we must ask ourselves: Are we creating tools that understand us, or mirrors that simply reflect our own biases and limitations?*

— JAMES MANYIKA ET AL.

Let's now discuss AI and social psychology. As we tread along and move past our discussion of human and computer interaction, I believe we could build upon that knowledge and color our understanding of psychology.

The role of AI in social psychology doesn't begin (and end) at mere observation; it has the potential to analyze vast amounts of data to reveal patterns in behavior and attitudes that we may not be consciously aware of.

As these technologies evolve, they could help us better understand the dynamics of group behavior, the formation of social norms, and the intricacies of all our interpersonal relationships, ultimately informing how we engage with one another.

SOCIAL PERCEPTION AND AI

As we discuss societal perception and AI, a meaningful concept that arises is anthropomorphism. This term refers to the act of attributing human traits, emotions, or intentions to nonhuman entities, such as animals, objects, or machines, particularly in the context of technology and AI.

Humanlike AI

People often find it easier to relate to an AI system or robotic entity when it exhibits humanlike characteristics.

For instance, in the series *Marvel's Agents of S.H.I.E.L.D.*, an AI system aptly called AIDA (which stands for Artificial Intelligence Digital Assistant) serves as a compelling example of anthropomorphism in AI. In it, AIDA is developed as an advanced AI capable of learning and adapting, possessing a humanlike personality and emotional expressions.

Her interactions with human characters highlight the tendency of individuals to attribute human traits to her, perceiving her as more than just a program. While this story with AIDA is fictitious, it's meant to serve merely as a concept to help you understand the point of anthropomorphism. Anthropomorphism has actually been a longstanding concept, even way back in the 1960s.

Four Degrees of Anthropomorphism

According to research that looks into the degrees of anthropomorphism, one of the earliest and most notable examples of anthropomorphism in computer science is ELIZA, developed at MIT in the mid-1960s by Joseph Weizenbaum (Gibbons et al., 2023). ELIZA was an early NLP program that simulated conversation by mimicking humanlike responses to user input, engaging users in a way that made them feel they were interacting with a real person.

This phenomenon of imbuing machines with humanlike qualities showcased two things: the potential of AI, and a reflection of a societal tendency to perceive technology through a human lens.

There are four degrees of anthropomorphism:

Degree of Anthropomorphism	Description	Examples
Basic anthropomorphism (courtesy)	This involves attributing simple human traits, like friendliness or approachability, to machines.	A friendly robot that greets users.
Advanced anthropomorphism (reinforcement)	This degree includes giving machines more complex humanlike characteristics, such as emotions or personality traits.	A virtual assistant that shows kindness and empathy.
Complex anthropomorphism (roleplay)	At this level, machines are viewed as having social roles or relationships similar to humans, including companionship or mentorship.	A robot that serves as a personal fitness trainer.
Intense anthropomorphism (companionship)	This is the highest degree, where machines are believed to possess full consciousness or self-awareness, leading to deep emotional bonds and humanlike interactions.	A highly advanced AI capable of deep conversations and emotional support.

Trust and Acceptance

Understanding anthropomorphism is important because it helps you give humanlike qualities to nonhuman characters, making them easier to relate to for readers. This creates stronger emotional ties, improves storytelling, and lets us explore complicated ideas like identity and existence.

As always, when dealing with AI, it's important to create a space where users feel valued. Many people may find interacting with AI challenging. Clear information about what the AI can do, its limitations, and how it can help users matters.

It's also important that users feel they have control over their interactions. People like to feel in charge of their tools, so letting them share their thoughts can help build confidence and trust.

The Role of Trust in AI

Trust is a critical factor in how people accept and engage with AI. If a user does not trust the AI, they are unlikely to rely on it for assistance or decision-making. Building trust involves several factors.

Transparency is key. If users understand how the AI works, what data it uses, and how it makes decisions, they are more likely to feel comfortable using it.

For instance, an AI that explains its reasoning when providing a recommendation will create a sense of trust.

Additionally, reliability is very important. Users need to see that AI systems work well and consistently. If an AI gives wrong answers or doesn't perform well in key situations, users may not trust it again. To improve reliability, developers should focus on thorough testing and ongoing improvements.

Acceptance of AI Systems

Acceptance of AI systems is tied to trust. Even if a user sees what AI can do, they may still be unsure about using it if they think it could threaten their job or skills. This is particularly true at work, where AI might be set up to help or even take the place of human workers.

For example, a team member might feel uneasy about using AI tools while fearing they might lose their job to the machine. Here, effective communication about the purpose of AI is very important. Organizations should clarify that AI is meant to be a tool that enhances human efforts, not replace them.

Also, providing user-friendly interfaces can significantly aid acceptance. Users are more likely to embrace AI if they find it easy to use. Simplified designs, clear instructions, and responsive support can help guide new users through their experience with AI systems, gradually increasing their comfort and acceptance.

The Interaction Between Trust and Acceptance

The relationship between trust and acceptance of AI is circular. Increased trust can lead to higher acceptance, while greater acceptance can foster trust. This dynamic is vital to understand, especially for developers and organizations looking to introduce AI technologies.

So, if users find an AI system helpful and user-friendly, their trust will likely grow. In turn, as they trust the AI, they may be more inclined to explore its features and capabilities.

Take self-driving cars developed by Tesla, for example. As users gain experience and see the safety measures in place, they might start to trust the technology more and subsequently accept it into their lives.

That said, here are the factors that influence trust and acceptance:

- **Accuracy:** The performance and reliability of AI outputs play a significant role. If AI consistently provides accurate and relevant results, trust is enhanced.
- **Communication:** Clear communication regarding AI capabilities and limitations helps set realistic expectations and builds trust.
- **Cultural factors:** Different cultures may have varying perceptions of technology and trust, influencing how AI is received.
- **Ethical considerations:** The ethical implications of AI, including fairness, bias, and privacy concerns, can either build or erode trust. Responsible design practices are essential.
- **Human oversight:** The presence of human oversight in AI operations can increase trust. Users are more likely to trust AI when they know humans are in control.
- **Regulation and standards:** Guidelines and policies governing AI use can create a framework that enhances trust. Compliance with standards can reassure users.
- **Reputation:** The credibility of the organization behind the AI influences trust. Established companies with a history of ethical practices tend to engender more trust.
- **Security:** Ensuring the security of data processed by AI systems is vital. Users need to feel their information is safe and protected.
- **Transparency:** The ability to understand how AI systems make decisions is important. If users can see the reasoning behind AI actions, it increases trust.
- **UX:** A positive interaction with AI systems fosters trust. This includes intuitive interfaces and responsiveness to user needs.

AI IN SOCIAL MEDIA AND COMMUNICATION

AI has become a key part of the online world, especially in suggesting content and checking what gets posted. Content recommendation means offering users items like articles, videos, or products based on what they like. Meanwhile, moderation is about reviewing content to make sure it follows community rules. AI greatly improves both of these areas.

For example, let's shed some light on Twitch, a 2011-founded streaming platform. It uses the power of AI to recommend streams and channels to viewers based on their watch history and preferences.

AI-Driven Social Media

AI programs look at how you act online to suggest content you'll like. When you watch a video on a streaming site, the AI checks your watching habits. It sees what types of videos you like, how long you stay on specific videos, and what other people who watch like you enjoy. From this information, the system can recommend new videos that you might like.

If you frequently watch documentaries about nature, the AI system will recommend similar content. And, if you often watch documentaries about serial killers, you might get suggestions for shows like *How to Get Away With Murder*.

Another aspect of AI in content recommendation is collaborative filtering. This technique looks at the preferences of many users to find patterns.

For instance, if User A and User B both watched similar content, the AI might suggest to User A content that User B found interesting. This cross-referencing allows the system to expand its recommendations beyond a single user's behavior, bringing a wider array of content into consideration.

The Importance of Content Moderation

Moderation is another area where AI plays a crucial role. The vast amount of content generated by users daily makes it challenging for human moderators to review everything. AI can automatically flag inappropriate content, such as hate speech, graphic violence, or explicit material, allowing human moderators to focus on more nuanced decisions.

For example, if a video posted by a user contains violent imagery, the AI can recognize this through image analysis. It uses patterns learned from existing databases to assess whether the content violates community guidelines. If it does, the AI can flag it for review or remove it automatically, providing a safer environment for users.

Building Better UX With AI

The blend of recommendation and moderation facilitated by AI creates a better overall UX. When users receive content that aligns with their interests, they are more likely to stay on the platform longer. At the same time, having a moderated environment makes users feel safer and more comfortable engaging with content.

Therefore, companies investing in AI technologies for both recommendation and moderation are setting themselves up for success. Users appreciate platforms that understand their needs and protect them from harmful content.

Psychological Impact

When we talk about the positive psychological impact, we are leaning toward the good things that can happen when our mental well-being is fostered and nurtured. This impact can manifest in various areas of our lives, including our relationships, work, and overall happiness. Understanding this can help us make choices that enhance our mental health.

Positive Psychological Impact

One way to create a positive psychological impact is through gratitude. Gratitude is the practice of acknowledging what we appreciate in our lives.

This simple exercise can shift your focus from what is lacking in your life to what is abundant. Taking the time to reflect on the positive aspects of our lives can greatly enhance our mental state.

Here are the positive psychological effects of AI:

Positive Psychological Effect	Description	Examples
Enhanced accessibility	AI can provide support to individuals with disabilities, boosting self-esteem and independence.	Voice-activated assistants help visually impaired individuals navigate.
Better decision-making	AI can analyze large amounts of data and provide insights for informed decisions in various aspects of life.	Personal finance apps analyze spending patterns to offer budgeting advice.
Personalized learning experiences	AI can create tailored educational content based on individual learning styles and paces, improving outcomes.	Adaptive learning platforms can adjust the level of difficulty depending on learning style.
Increased creativity	AI tools can assist artists and creators by providing inspiration and enhancing creative possibilities.	AI programs generate music or art that artists can build upon.
Data-driven insights	AI can analyze behavior and preferences, helping users gain self-awareness and personal development.	Fitness apps provide personalized workout suggestions based on user data.

Efficiency in daily tasks	By automating routine tasks, AI reduces cognitive load and stress, allowing focus on more meaningful activities.	Smart home devices automate routine chores like cleaning and temperature control.
Crisis management	AI can aid in disaster response by analyzing data and improving communication, enhancing safety and preparedness.	AI systems can predict natural disasters and provide timely alerts to communities.

Negative Psychological Impact

To mitigate the negative psychological effects of AI, it is vital to cultivate a positive relationship with technology. Learning to use AI in a way that enhances rather than detracts from our well-being is essential.

Here are the negative psychological effects of AI:

Negative Psychological Effect	Description	Examples
Bias and discrimination	Experiencing biased AI systems can cause frustration, helplessness, and mistrust among individuals and groups.	AI hiring algorithms favor certain demographics over others.
Dependence on technology	Relying too much on AI can reduce critical thinking, problem-solving skills, and decision-making abilities.	Overreliance on GPS for navigation leads to difficulty with spatial awareness.
Erosion of privacy	Continuous surveillance and data collection through AI can lead to anxiety about personal privacy and a sense of losing control over personal information.	Concerns over personal data being sold or misused by apps.
Impaired emotional processing	Relying on AI for emotional support can hinder the development of healthy coping mechanisms and emotional intelligence.	Using chatbots for therapy instead of seeking real human help.
Job anxiety and insecurity	Fear of losing jobs to automation can create stress and anxiety, impacting job satisfaction and mental well-being.	Workers in manufacturing fear layoffs due to robotic automation.

Manipulation and misinformation	AI-generated false information can create confusion, distrust in information sources, and anxiety about discerning truth from lies.	Deepfake videos cause skepticism about authentic media content.
Reduced social interaction	Using AI chatbots or virtual assistants may lead to less face-to-face communication, causing feelings of loneliness and isolation.	Substituting conversations with virtual assistants for socializing with friends.
Identity and self-worth issues	Interacting with AI may challenge how individuals see themselves and impact feelings of self-worth, especially when compared to AI outcomes.	Users feel inadequate when comparing their skills to AI-generated performances.

IMPACT OF AI ON SOCIAL BEHAVIOR AND INTERACTIONS

The way we act in social situations has changed a lot because of AI. AI is now a part of many things we do every day, like using social media and personal devices. As we use AI more in our interactions, our behavior and what we expect from others also change.

AI in Collaborative Work

AI affects teamwork by making communication easier. AI tools can summarize meetings, point out important information, and help with scheduling tasks.

For example, tools like Otter.ai and other AI meeting assistants can transcribe conversations and provide key takeaways after discussions. This allows team members to focus on the content of the meeting rather than taking notes. With this technology, everyone stays informed and aligned on project goals.

Building Stronger Connections

Even in a virtual environment, AI can help build stronger connections among team members. Social collaboration platforms often include features that facilitate team bonding.

AI can suggest activities for team building or create spaces for casual conversations. This helps team members build friendship and trust, which is important for working well together. When teams feel connected, they communicate better and get more done.

AI and Social Norms

As AI evolves, it's essential to consider the social norms surrounding its use. They refer to the accepted behaviors and attitudes within a society or group. Examples include polite greetings, punctuality, dress codes for different settings, appropriate levels of physical contact, and expectations regarding family roles and responsibilities.

Social norms dictate how individuals interact with each other and technology. And, in many ways, AI can shape these norms, just as they can influence how AI is developed and employed. Understanding the social norms related to AI involves recognizing that people are deeply influenced by the technology they use.

A TWO-WAY RELATIONSHIP

In the emerging highly programmed landscape ahead, you will either create the software or you will be the software. It's really that simple: Program, or be programmed.

— DOUGLAS RUSHKOF

We can't talk about AI without talking about psychology. With every job we ask AI to do for us, we're asking it to replicate human thinking and behavior in some way, and that would be impossible without looking at human psychology... Yet its significance is rarely brought up in public discourse about AI, and this is something we need to be aware of because it's not just human psychology that affects AI—it works the other way too. We need to be aware of what we're working with and how it affects us: It's like Douglas Rushkoff said: "Program or be programmed."

This isn't just the case with AI, of course: It's something we should be aware of with every new development that becomes so widely absorbed and changes the way we operate. We need to understand these things so that we understand ourselves and how our behaviors are influenced. I wrote this book because I want to open the gateway to understanding for more people, and now that you're along for the ride, I'd like to ask for your help in doing that. All you have to do is leave a short review online.

By leaving a review of this book on Amazon, you'll make it more visible to new readers who are looking to broaden their understanding of the relationship between AI and psychology.

Reviews make it easier for people to find the information they're looking for and let them know how other readers have experienced the book, so while it may not seem like much to you, a few words from you will make a real difference in getting this information out to more people. AI is here to stay, and the future is exciting—but to make the most of it, we need to understand how it intersects with psychology, and we need to understand the two-way relationship we have with it.

Thank you so much for your support. Now, let's get back to business!

Scan the QR code below

ETHICAL AND PSYCHOLOGICAL
IMPLICATIONS OF AI

I hope you enjoyed our discussions in the previous chapter about AI and its psychological implications. I also hope you learned what you needed to know because our discussions were somewhat of a jumping-off point for this chapter's topic.

And, to start this chapter, I believe a concept about the ethical and psychological implications of AI is the three-body problem—an important idea in physics and astronomy.

Unlike the simpler two-body problem, which is easily solvable and allows us to predict the orbits of two objects around each other, the three-body problem is much more complicated. When we add a third body, the calculations become unpredictable and complex.

Let's look at a real example involving AI, its users, and the ethical issues these interactions raise. We can see the three-body problem as a way to understand the complicated relationships among AI systems, their users, and the values of society.

Just like the way three planets can move in unexpected ways because of their gravitational pull, the relationship between what AI can do, what users want, and ethical standards can lead to complex and hard-to-predict results.

As this interaction changes, being open and accountable becomes very important for ethics. Users need to be aware of how they can affect AI systems, and developers should think about the ethical impact of their designs. The three-body problem helps us look at how to use AI responsibly in society and shows the need for working together.

ETHICAL CONSIDERATIONS IN AI DEVELOPMENT

Addressing bias in AI algorithms is a critical task for many professionals in technology today. Bias can occur when the data used to train these algorithms is not representative of the real world.

Bias and Fairness

To better understand how to address bias in AI, we can break the process down into several main steps.

Here are the different types of biases:

Type of Bias	Description	Specific examples
Confirmation bias	AI focuses on data that supports its existing patterns and ignores data that contradicts them.	An AI model trained on biased data continues to reinforce stereotypes.
Anchoring bias	AI gives too much weight to the first data it processes when making predictions or decisions.	An AI model adjusts predictions based mainly on initial training data.
Hindsight bias	AI believes it could have predicted outcomes after they occur.	After a market crash, an AI claims it would have detected the signs.

Self-serving bias	AI highlights its successful predictions while downplaying or ignoring its errors.	An AI highlights high accuracy rates while ignoring misclassified cases.
Groupthink	AI systems influenced by similar algorithms may converge on the same solutions.	Different AI models provide the same flawed recommendation.
Availability bias	AI prioritizes findings or data that are easily accessible.	An AI model relies heavily on popular data sources, neglecting rare cases.
Overconfidence bias	AI becomes overly confident in its predictions.	An AI claims a 95% accuracy rate without validating against new data.
Status quo bias	AI may resist changes in its algorithms or data inputs.	An AI refuses to adapt its model despite new and improved algorithms.
Bandwagon effect	AI replicates popular trends in data and decisions.	An AI model adopts a widely used but flawed algorithm without scrutiny.

How to Be Fair

It's important to be fair in our decisions, even when biases can affect us. Studies have found that people who treat others fairly can achieve much better results.

In light of this, you can say the same about being fair when working with AI systems.

Here's what you can do to be fair:

- **Conduct impact assessments:** Before deploying AI systems, assess their potential impact on various groups, ensuring that no group is unfairly disadvantaged.
- **Diversify training data:** Ensure that the data used to train AI models comes from a wide range of sources and demographic groups, reducing bias.
- **Engage in collaborative development:** Work with organizations and communities affected by AI decisions to

cocreate solutions, ensuring their needs and perspectives are integrated.

- **Incorporate AI for good initiatives:** Actively engage in projects that use AI to address social issues, ensuring that technological advancements are aligned with societal benefits.
- **Promote ethical AI communities:** Foster and participate in communities focused on ethical AI practices, sharing knowledge and best practices to advance fairness collectively.
- **Regularly audit algorithms:** Conduct frequent audits of AI systems to identify and address biases, ensuring that decisions remain fair over time.
- **Use fairness metrics:** Implement specific metrics to measure fairness in AI outputs, adjusting algorithms as needed to improve fairness outcomes.

Accountability and Transparency

Accountability is important in decision-making. Being accountable means taking responsibility for what you do and the results that come from it.

From a technical standpoint, this means that developers and organizations must take responsibility for the behavior and outcomes produced by their algorithms. They have to ensure people understand how AI systems make decisions and be aware of possible biases from the data used to train these systems.

When you make a decision, big or small, being accountable means you're ready to deal with the outcomes. It's not just about admitting what you did, but also seeing how your choices affect other people and the world around you.

PSYCHOLOGICAL IMPACT OF AI ON EMPLOYMENT AND SOCIETY

We often hear about the technological advancements that AI has made. Although, it's important to understand how these advancements impact our psychological health and societal structures.

For many people, the introduction of AI in workplaces can lead to feelings of anxiety and uncertainty about job security and the future of their roles. As companies increasingly automate tasks, employees may fear that their skills will become obsolete or that they will be replaced by advanced technology.

Job Displacement and Transformation

The rise of AI is a double-edged sword. And, inarguably, it has changed many aspects of how we live and work.

Here's a look into the contradictory characteristics of AI:

- **Efficiency vs. job loss:** AI can make things work faster, but it might take away jobs from people.
- **Better decisions vs. bias:** AI can help make smarter choices, but if trained on unfair data, it can treat some people unfairly.
- **Innovation vs. ethics:** AI can create cool new things, but it also raises worries about privacy and how it can be misused.
- **Personalization vs. privacy:** AI can give us custom experiences, but it needs a lot of our data, which can lead to privacy issues.
- **Competition vs. tensions:** Countries and companies want to be the best at AI, which can lead to progress but might

also create conflicts between those with and without access to technology.

Adapting to Change

The pressure of losing jobs to AI raises important questions about how individuals and society can adapt to these changes. Workers need to invest in their education and skills.

Upskilling, or learning new skills that are more relevant in the job market, can help individuals remain competitive. This might involve pursuing further education, attending workshops, or even taking online courses. A recent survey found that over 70% of employees believe upskilling is essential for their career advancement and job security (Brower, 2022).

One helpful approach is to focus on skills that AI cannot easily replicate.

Here are the skills you need:

Skill	Explanation	Example
Emotional intelligence	The ability to recognize, understand, and manage emotions in oneself and others, fostering empathy and interpersonal connections.	A manager resolves a team conflict by understanding the feelings of each member.
Creativity	The capability to generate original ideas, concepts, or solutions that are imaginative and innovative, often requiring human intuition and storytelling.	An artist creates a unique piece of art that tells a personal story.
Critical thinking	The skill to analyze information, evaluate arguments, and make informed decisions, integrating context and human experience.	A scientist designs an experiment based on carefully evaluated data.
Intuition	The ability to understand something immediately, without the need for conscious reasoning, drawing on personal insights and experiences.	A chef senses that a dish needs more seasoning without measuring.

Moral judgment	The capacity to make ethical decisions based on complex societal values, feelings, and cultural understandings, which involve subjective reasoning and beliefs.	A lawyer chooses to defend a client based on a belief in justice, despite societal bias.

Becoming Versatile

Another way to prepare for the job market's evolution is by becoming versatile. Diversifying one's skill set can open up new opportunities.

For example, someone with a background in marketing can learn about data analytics to enhance their employability. They can also explore digital marketing strategies that leverage AI tools or have an excellent grasp of cybersecurity.

Societal Change

AI's influence is evident in how values shift over time. With the rise of AI-generated content, there is a growing debate on the importance of originality. This brings about a conversation regarding creativity and authenticity.

For instance, if an AI can produce music that sounds indistinguishable from that of a human artist, where does that leave traditional musicians? The values associated with creativity, expertise, and hard work could evolve dramatically as AI becomes more prevalent in various artistic domains.

AI AND PRIVACY CONCERNS

Many companies use AI to manage large amounts of personal data, with a study showing that 80% of organizations use AI technologies to enhance their data management capabilities (Vena Solutions, 2024).

For instance, Salesforce, a cloud-based software company, utilizes AI through its Einstein platform to analyze customer data and provide personalized insights. This brings up important questions about how this information is used, stored, and kept safe.

Data Privacy

It's important to understand what personal data is. Personal data includes names, email addresses, phone numbers, or even biometric data like fingerprints. Awareness of what counts as personal data helps in handling and protecting this information properly.

A simple way to get around this is to use data minimization techniques. Data minimization involves reducing the data collected and retaining only what is necessary.

Here are some data minimization techniques to be aware of:

- **Access controls:** Restrict access to the data to only those individuals or systems that require it for the intended purpose.
- **Data collection limitation:** Collect only the essential data for fulfilling the specific purpose. Avoid collecting extra information that isn't needed.
- **Encryption:** Use data encryption techniques to protect sensitive information at rest and in transit, minimizing the risk of unauthorized access.
- **Purpose limitation:** Ensure that data is collected only for legitimate purposes and not used for unrelated activities.

- **Regular audits and assessments:** Conduct periodic reviews of data handling practices to ensure compliance with data minimization principles and identify areas for improvement.

AI Surveillance and Consent

AI surveillance is becoming more common in our daily lives. This raises many ethical questions that can impact our privacy and freedoms. It's important to be aware of these issues to better understand the world we live in today.

Invasion of Privacy

One big ethical problem with AI surveillance is that it violates people's privacy. When these systems are in place, many people aren't aware that someone is watching them. This can cause feelings of unease or fear, knowing that their actions might be tracked.

For example, CCTV cameras equipped with AI can recognize faces, track individuals, and even analyze their behavior. While these technologies may help reduce crime, they can also invade the privacy of innocent individuals who are merely going about their daily lives.

Data Collection and Consent

Another ethical concern is the collection of data without explicit consent. Many AI surveillance systems gather vast amounts of data from the public, often without informing those being monitored.

One notable case is the use of Clearview AI. This facial recognition technology company has gained significant attention for scraping images from social media and other public websites to create a massive database of faces. Law enforcement agencies have used this technology to identify suspects in various investigations.

However, this has raised serious ethical questions, as individuals whose images were collected did not know that their likenesses were being used for surveillance purposes. Critics argue that this practice infringes on privacy rights and lacks transparency, as the subjects of the data collection were not informed or asked for consent to have their images stored and analyzed.

This can include video footage, social media activity, and even location tracking through mobile devices. The absence of clear consent raises questions about who owns this data and how it can be used.

False Positives and Errors

AI systems are not perfect, and their accuracy can sometimes lead to serious ethical implications. AI-driven surveillance can produce false positives, meaning that individuals may be wrongly identified as suspects or criminals.

For instance, if a surveillance system misidentifies someone based on their facial features or behavior, that person may face unwarranted scrutiny from law enforcement or the public. This raises fundamental questions about accountability and who to blame when these technologies fail.

Chilling Effect on Freedom

The presence of AI surveillance can have a chilling effect on people's freedom of expression and assembly. When individuals know they are being monitored, they might be less likely to express their opinions, participate in public protests, or engage in other forms of social interaction. This can stifle dissent and hinder democratic processes.

For instance, the activism surrounding Black Lives Matter (BLM) has faced scrutiny from law enforcement agencies. During one

BLM rally, the use of advanced surveillance technologies to monitor protests and gatherings was received poorly.

When people know they are being watched, they may hesitate to share their thoughts, join protests, or interact with others. This can reduce disagreement and affect democratic activities. Activists may feel scared about the possible consequences of their involvement in these movements.

AI IN THERAPY AND MENTAL HEALTH

In mental health, AI is making its presence apparent, too. Now that we've completed our discussion about AI's ethical and psychological implications, let's go into mental health.

As a way of introducing this chapter, I'll talk to you about Ginger, an on-demand mental health platform based in California. Ginger presents a range of services, including chat-based coaching, self-care resources, and video therapy to people who need help with their mental health.

In 2021, Ginger merged with Headspace, a California-based company that makes meditation and mindfulness accessible, in a $3 billion deal (Forrester, 2021).

I believe this was a good merger because it made the need to seek help less challenging. According to a Statista study, 56% of adults who were surveyed needed help (Vancar, 2023).

But, while this many people needed help, the sad fact remains that they didn't ask for it.

So, in light of this, let's talk more about the other fascinating advancements in therapy and mental health. And, this chapter is the perfect venue for this.

AI-DRIVEN THERAPEUTIC INTERVENTIONS

AI is becoming more important in therapy and counseling. According to a report, the global market for AI in healthcare is projected to reach $34 billion by 2025, indicating a growing reliance on technology-driven solutions (Brown, 2018).

This increase shows that more people see how AI can improve mental health services. It offers tools for personalized therapy, real-time monitoring, and easier access for those who need help. By using AI in therapy, mental health professionals can use data to create better treatment plans, which leads to improved patient results and a more effective healthcare system.

Virtual Therapists and Chatbots

One of the main advantages of AI in therapy is accessibility. Many people live in areas where professional therapists are not available. They might also have busy schedules that make it difficult to attend regular therapy sessions.

Connecting Users With Human Therapists

While AI tools are helpful, they do not replace the human touch in therapy. If a chatbot recognizes that a user is experiencing severe distress, it can recommend that they seek help from a professional. This connection is vital because even though AI can offer support, it is not equipped to handle every situation.

An example is Woebot Health. It's a company that specializes in mental health technology, particularly through the use of conversational agents or chatbots.

Their primary product, Woebot, is an AI-powered chatbot designed to provide support and guidance for mental well-being. The chatbot uses cognitive behavioral therapy (CBT) principles to engage users in discussions about their feelings, thoughts, and behaviors.

Woebot is designed to be available 24/7, which is why it can offer users an anonymous and accessible way to talk about their mental health. It can help users track their mood, provide coping strategies, and deliver psychoeducation to help them better understand their mental health challenges.

Overcoming Stigma With AI Support

The stigma surrounding mental health can prevent many individuals from seeking help. AI tools can help break down some of these barriers. Because they provide anonymous support, people might feel more comfortable expressing their thoughts and emotions without fear of being judged.

AI IN DIAGNOSTICS

One of the most interesting applications of AI is its potential use in diagnosing mental health conditions. This could change how we approach mental healthcare and make it more efficient and accessible for everyone.

How AI Can Help Diagnose Mental Health Conditions

AI can analyze large amounts of data much faster than a human can. When it comes to mental health, this means it can sift through a multitude of patient records, surveys, and even social media interactions to identify patterns that may indicate a mental health issue.

For example, if someone posts frequently about feeling anxious or depressed, an AI system can recognize these trends over time. This kind of analysis can help mental health professionals understand a person's condition better and provide a more tailored approach to treatment.

The Importance of Psychometric Tools

Psychometric tests, which measure mental abilities and behavioral style, are vital in mental health. AI can enhance these tools by providing a more objective assessment. For example, an AI-driven application could administer questionnaires that adapt based on previous answers.

For one, the Perceived Stress Scale (PSS) is a popular tool for measuring how stressed people feel. Sheldon Cohen created it to help evaluate how unpredictable, uncontrollable, and over-whelmed people feel in their daily lives.

The PSS works simply. It focuses on the subjective nature or feelings of stress rather than the events themselves.

CHATBOTS

Like Woebot, other mental health chatbots use computer programs designed to simulate a conversation with users. These chatbots are specifically geared toward helping individuals who may be struggling with mental health issues.

Design and Functionality

Effective design in mental health chatbots is important because it affects how users feel about using them. These chatbots provide support, useful information, and a safe space for users to express their emotions. Creating a good mental health chatbot requires careful thought and planning.

Understanding the User's Needs

The first step in creating a successful mental health chatbot is understanding the user's needs. This involves recognizing what individuals might be looking for: immediate assistance, emotion tracking, or accessing helpful information.

For example, someone feeling anxious might want quick breathing exercises or reminders to take breaks.

Designing Conversational Flow

Creating an effective conversational flow is critical for user engagement. The chatbot should be able to guide users through various topics without creating confusion.

For example, if a user mentions feeling sad, the chatbot can respond by asking if they want to talk about it or provide coping strategies. Mapping out different paths for conversations helps ensure users do not feel lost or frustrated. Tools like flowcharts can be useful in visualizing how conversations may progress based on user inputs.

Implementing Friendly and Supportive Language

The language used in a mental health chatbot is crucial. It should be friendly, supportive, and nonjudgmental. Simple language helps users feel comfortable sharing their thoughts.

For example, instead of saying, "You shouldn't feel that way," the chatbot can reply, "It's okay to feel that way; many people do." This small change can make a huge difference in how someone perceives the conversation. Ensuring that responses remain positive yet realistic encourages users to continue interacting with the chatbot.

Providing Crisis Resources

While chatbots can offer valuable support, they cannot replace human help in dire situations. Including clear information on crisis resources is critical. This might involve directing users to hotlines or emergency contacts they can turn to in moments of crisis.

For example, if a user expresses thoughts of self-harm, the chatbot could immediately provide a contact for a crisis hotline, ensuring that the user is directed toward immediate help.

Efficacy and Acceptance

When we talk about evaluating a chatbot, we need to understand what efficacy means in this context. Efficacy refers to how well the chatbot performs its intended functions. In other words, we want to see if the chatbot can provide accurate information, handle user inquiries effectively, and enhance the overall UX.

Response Accuracy

We should assess the accuracy of the chatbot's responses. When users ask questions, they expect precise answers. If the chatbot provides incorrect or vague information, it can harm the user's trust in the service.

One way to evaluate response accuracy is by comparing the chatbot's answers to a set of verified answers.

For instance, a customer asking about store hours expects a specific timeframe. If the chatbot says, "We're open sometime in the afternoon," that is not satisfactory. Tracking the percentage of accurate responses over time can help measure how many correct answers the chatbot provides.

User Satisfaction

User satisfaction is another critical component of evaluating a chatbot's effectiveness. Surveys can be a useful tool for measuring how users feel about their experiences with the chatbot.

After interacting with the bot, users can be asked to rate their satisfaction on a scale from 1 to 5. This feedback can provide insight into what users appreciate about the chatbot and what improvements are necessary.

Handling Complexity

Chatbots need to be able to handle complex inquiries as well. A well-functioning chatbot doesn't just respond to simple questions; it should also manage more nuanced scenarios.

For example, if a user has a problem with an order, the chatbot should ask questions to understand the situation better. It might need to gather specific details, such as the order number or the nature of the issue.

Evaluating how well a chatbot handles these scenarios can provide valuable insights into its capabilities. Success in this area can often be enhanced through continuous learning and updates in the chatbot's programming, enabling it to tackle increasingly complex questions.

Response Time

Response time is a significant factor in user satisfaction as well. Users expect quick responses, so evaluating how fast the chatbot replies when a question is asked is crucial. Delays in response time can frustrate users and lead to a poor experience.

Measuring the average time taken for responses can help organizations identify lags in performance. If a chatbot takes too long to reply, it may indicate the need for optimization, ensuring that users receive answers promptly.

Context Awareness

This means the chatbot should not only recognize the individual words users type but also grasp the overall meaning of a conversation. Evaluating a chatbot's context awareness can involve complex testing scenarios, where varied topics and questions are thrown at the bot to see if it remains aligned with the conversation.

For example, if a user is discussing a technical issue and later asks about related troubleshooting steps, the chatbot should understand the context instead of treating the new question as completely separate.

AI FOR MENTAL HEALTH MONITORING AND SUPPORT

AI tools can analyze your data and provide insights that might not be obvious. They can detect changes in your mood or behavior over time, offering a clearer picture of your mental state. For example, some apps allow users to log their emotions and activities, using algorithms to provide personalized feedback. These apps can suggest coping strategies based on the information you provide.

Continuous Monitoring

Tracking your mental health is important for feeling good overall. It helps you see how your feelings and actions change over time. By keeping an eye on your mood, you can spot signs of stress, anxiety, or sadness. This knowledge lets you take steps to handle these feelings better.

There are various AI-driven apps and platforms available to assist with mental health tracking. Popular options are Moodtrack Diary, Daylio, and other mood-tracking applications. These apps allow users to input their daily mood, jot down any triggers, and note their activities.

Over time, the app will provide insights based on the input data. Some platforms even incorporate AI chatbots that can converse with users about their feelings, giving immediate support. Others may offer meditation or relaxation strategies tailored to your recorded moods.

Tips for Effective Mental Health Tracking

To get the most out of tracking your mental health with AI tools, there are some tips to consider. If you find it difficult to identify your emotions, consider using prompts or questions to guide you.

Recognizing Patterns and Making Changes

As you track your mental health over time, pay attention to the specific patterns that emerge in your mood, energy levels, and stress triggers. Conversely, neglecting to monitor these aspects can lead to a lack of awareness about your emotional well-being, causing you to overlook significant changes in your mental health.

For instance, consistently noting your feelings in a journal or using a mental health app can help you pinpoint fluctuations tied to certain events, situations, or interactions. The more you engage in this reflective process, the deeper awareness you'll develop regarding your emotional landscape and how external factors influence it.

You may start to recognize that your stress notably increases during work deadlines, particularly on days when multiple deadlines coincide or when you feel underprepared. Once you identify this connection, you might explore specific techniques such as structured time management strategies, prioritization of tasks, or even mindfulness practices to ease that stress and enhance your productivity during these peak periods.

SUPPORT SYSTEMS

Support systems serve as a safety net. When life throws curveballs —like job loss, relationship breakups, or health issues—having people around you can help you cope. It's important to recognize that everyone faces tough times, and being able to talk about these issues can reduce feelings of loneliness.

Emotional Support

One main aspect of a support system is emotional support. This kind of encouragement allows people to feel understood and cared for. Let's take the common situation of going through a rough patch.

Talking to a friend about how you feel can lift some weight off your shoulders. They may offer comforting words or remind you of your strengths. This emotional bonding offers a sense of belonging and community.

Practical Assistance

Support systems also provide practical assistance. Sometimes, you just need someone to help you juggle tasks or tackle challenges head-on.

Let's say a friend might help you move, or a family member could pitch in for grocery shopping during a busy week. These small yet meaningful gestures can ease your burdens significantly.

If you think about it, many people have been in situations where they had too much on their plate. Having someone step in to help can be a game-changer.

For example, when a new parent is adjusting to the demands of a baby, having family or friends around to help with meals or babysitting can make the transition smoother.

Maintaining Your Support System

A support system needs care and attention. Spend time on your relationships. This could mean having dinner with friends every month or calling family members to see how they are doing. The more time you put in, the stronger these connections will be.

It's also vital to be there for your support system. Feel free to offer help when they need it, too. This reciprocal relationship strengthens bonds and creates a lasting network of caring individuals.

AI AND HUMAN CREATIVITY

There's a running argument about how AI is impacting creativity. Some believe that AI is diminishing creativity, while others think it enhances it.

What do you think?

One viewpoint is that AI tools are taking away the uniqueness of creative work. Many artists and writers express fear that their individual touch might be lost when machines can replicate styles and generate pieces at a much faster rate.

However, looking at the other side, AI can actually complement human creativity. It can offer suggestions or generate ideas that a person might not have thought of otherwise.

For example, an author might use an AI tool to brainstorm story ideas or character traits. Instead of replacing the human element, AI can act as a helpful partner. It provides artists with new perspectives and can inspire them in unexpected ways.

Regardless, AI is here to stay. In the discussions below, let's talk about how you can use it accordingly. In Chapter 8, we talked about AI's role in mental health. Now, let's talk about its place in human creativity.

AI IN CREATIVE FIELDS: ART, MUSIC, AND WRITING

When AI enters the fields of art, music, and writing, it changes how these creative areas work by providing new tools and methods to improve the creative process. AI allows artists to try out new styles and ways of expressing themselves that weren't possible before, resulting in unique forms of art.

A McKinsey & Company (2024) report that provides data on the state of AI says that by 2030, 72% of companies are likely to use some kind of AI technology, which shows how much AI is affecting different industries, including the arts. This use of AI helps artists explore new ideas and create original works that push traditional limits.

Generative AI

Generative AI refers to systems that can create content based on input data. For instance, it might analyze existing pieces of artwork and generate new images that are similar yet unique. This ability often sparks lively debates among artists, musicians, writers, and audiences about the nature of creative work.

Understanding Generative AI

Generative AI uses programs to create new content. In art, this means it can make paintings or digital artwork by learning from many images. This technology often uses ML, which teaches the AI to recognize patterns and styles by studying large sets of data.

Here's a list of features that help you see the difference between work by generative AI and human-only work:

Generative AI Work	Human-Only Work
Creates content based on input data, analyzing patterns and styles from existing works.	Generates content driven by personal experiences, emotions, and cultural influences.
Generates new images similar to existing ones by processing a vast database of artworks.	Creates unique art pieces that reflect individual creativity and artistic vision.
Can produce music compositions by emulating styles from various genres.	Composes music that is deeply personal and often tells a story or conveys specific feelings.
Generates text based on patterns found in vast amounts of literature and other written works.	Writes narratives or poetry inspired by human experiences, thoughts, and emotions.
Operates quickly, producing multiple versions of a concept with minimal input.	Takes time for ideation, brainstorming, and iterative creation processes to refine ideas.
Lacks consciousness and emotional depth in its creations.	Infuses work with personal meaning, emotional resonance, and cultural context.
Uses algorithms and programming to create content with no subjective interpretation.	Relies on intuition, personal insight, and the unique human experience to shape creative output.

For example, an AI trained in various paintings from different eras may create an entirely new piece that blends elements from Impressionist and Abstract art.

When AI creates music, it starts by studying various musical compositions. It learns which notes are commonly played together and recognizes different styles. It can then compose a melody that fits within the typical confines of a certain genre, like classical or jazz.

Controversies Surrounding AI Creation

The arrival of generative AI in creative fields does not come without controversy. One major concern is whether AI-generated works can be considered "true" art, music, or writing.

What is creativity? Is it about why someone creates something, or is it just about the final result?

Many artists argue that human emotion and experience are vital components of creativity. They believe that AI cannot infuse genuine feelings into its creations. This can lead to significant discussions about the role of emotion and experience in various forms of expression.

Embracing AI as a Tool

Despite these concerns, many see potential in AI as a tool rather than a replacement for human creativity. Artists can harness AI to brainstorm ideas or explore new forms of expression they might not have considered.

For example, an artist might use AI to generate design ideas for sculptures, and then adapt those ideas into their unique vision. A specific company that exemplifies this approach is Artbreeder, which allows artists to blend and modify images using AI algorithms, leading to innovative design ideas. This way, AI can enhance the creative process rather than hinder it.

HUMANS AND AI COLLABORATION IN CREATIVE PROCESSES

When people think about AI, they often envision robots taking over jobs or creating things from scratch. But, in my opinion, their fear of AI taking over stems from the fact that they don't really know how to use AI.

Enhancing Human Creativity

Using AI as a tool means recognizing its strengths and combining them with human skills. It's not as simple as, "Do this for me," or,

"Do that for me." The writer remains in control, and the AI acts merely as a supportive partner in the creative process.

The AI takes the lead in generating ideas, often determining the direction of the narrative, while the writer merely edits and refines the content, relinquishing some control over the creative process.

For example, writers can use AI to find new ideas or generate stories. By inputting specific prompts, they can receive suggestions that spark inspiration. This way, the writer can take those ideas, develop them further, and shape them into something truly unique and personal.

The Importance of Human Input

It's important to remember that human ideas are vital in being creative. AI can look at data and come up with answers based on patterns, but it doesn't have the feelings and understanding that people do.

For instance, an artist may use AI to generate a background for a painting. However, the artist's personal touch, experiences, and feelings will guide how they adjust and refine that background to create a cohesive and meaningful piece of art.

You need to balance the quickness and efficiency of AI with human creativity and emotional understanding. When a designer uses AI to create prototypes, they can easily get different options and styles. The designer then picks the parts they like and blends them with their ideas. This teamwork encourages more creativity and exploration in design, leading to more unique results.

Limitations and Considerations

While AI is a powerful tool, it's also important to acknowledge that it's only that: a tool that has limitations. AI works based on patterns and data. It can't understand context deeply.

This means that sometimes proposals generated by AI may feel out of place or inappropriate. That's where human input enters the picture. Therefore, always consider whether the suggestions align with your project's goals and emotional tone.

Three Case Studies of Successful Human–AI Collaborations in Creativity

In recent years, the collaboration between humans and AI has yielded results that are nothing short of fascinating. These partnerships have pushed the boundaries of creativity, offering new ways to express ideas and create art.

Case Study 1: Film and Animation

In film and animation, the collaboration between humans and AI has led to innovative storytelling techniques. AI can assist in various aspects of production, from scriptwriting to post-production effects.

When creating an animated film, a team might use AI algorithms to assist in animating characters. Instead of manually creating every frame, AI can help generate movements based on input parameters. This significantly speeds up the animation process, allowing studios to focus more on crafting compelling narratives and expanding creative ideas.

For example, some filmmakers use AI to analyze audience preferences and gauge which elements resonate most effectively with viewers.

Specifically, in *Space Jam*, the blending of realism and animation is vividly showcased through Michael Jordan's character, who is a live-action NBA superstar, interacting with animated Looney Tunes characters like Bugs Bunny and Daffy Duck. The film was released in 1996.

For instance, during the basketball game, Michael Jordan dribbles the ball and performs realistic moves, while simultaneously passing the ball to Daffy Duck, who comically stretches and morphs in exaggerated ways typical of animation.

Case Study 2: Game Development

The field of video game development has also benefited from AI involvement. Developers can use AI to create dynamic characters and responsive gameplay environments. AI systems can analyze player behavior and adapt game challenges accordingly, creating a customized experience for each player.

For instance, an example of a video game that has effectively implemented AI is Final Fantasy VIII, developed by Square Enix. This role-playing game utilizes an advanced AI system to create a dynamic gameplay experience.

Case Study 3: Advertising and Marketing

Human–AI collaboration is also prominent in advertising and marketing. Advertisers use AI tools to analyze consumer data and tailor their campaigns effectively. By processing large datasets, AI can identify which types of messaging work best for different demographics.

For example, AI can help a marketing team divide its audience based on what they do online or what they buy. The AI can suggest content or ads that fit each group better. This helps companies connect with their audience more efficiently and get more value from their advertising spending.

PSYCHOLOGICAL PERSPECTIVES ON AI CREATIVITY

Many people believe that using AI to create art is unfair because it mimics the work of other artists. Using AI to make art raises questions about what makes art real and special. Some think that AI art is unfair because it copies other artists' styles, which takes away from the uniqueness and creativity that come from human artists.

In a piece published in the scientific journal *Sage Journals*, Walter Benjamin would say that when art is reproduced mechanically, it changes how people see and experience it (Kalpokas, 2023). This shift challenges our ideas about originality and the worth of creativity.

This viewpoint raises important questions about originality and authorship in the age of technology. When we think about imitation, we can break it down into several components that help us understand what it truly means in the context of AI and creativity.

Defining Imitation in Art

Imitation in art refers to the act of copying or replicating another artist's style, techniques, or ideas. This is not a new concept; throughout history, artists have looked to their predecessors for inspiration.

During the Renaissance, many artists studied and copied the works of great masters like Michelangelo and Leonardo da Vinci to create their own styles. Similarly, AI systems look at existing art to make new art. The main difference is in how AI and humans work.

The Risks of Overreliance on AI

On the other hand, there are inherent risks in relying too heavily on AI for creativity. If artists become overly dependent on AI systems, there may be a decline in traditional skills and techniques.

Here are discussions on what constitutes overreliance on AI:

- **Dependency on AI for social interaction:** Using AI for communication can weaken real-life relationships. People might prefer talking to AI instead of connecting with others, leading to feelings of loneliness.
- **Diminished creativity:** When artists and writers depend heavily on AI for ideas, it might limit their creativity. They may end up repeating what AI suggests instead of coming up with original work.
- **Loss of critical thinking skills:** Relying too much on AI for information can weaken our ability to think critically. People may stop questioning AI answers and just accept them without considering other options.
- **Reduced human judgment:** Relying too much on AI for decisions isn't wrong. But, it can lessen our confidence in our judgment. People may stop trusting their ability to solve problems or make ethical choices when they lean on AI too often.

AI AND HUMAN IDENTITY

Now that we've finished our discussion on AI and creativity, let's now go over discussions about AI and human identity.

This has boggled my mind because the way a person interacts with an AI system reflects their personality and their approach to technology. When you engage with AI, whether it's through voice commands, text, or another interface, your choices reveal your preferences.

For example, if you often ask detailed questions, it shows that you value thorough information and want to understand more. This can indicate that you are someone who enjoys learning and seeks to gather as much knowledge as possible before making a decision.

In contrast, if you tend to use short commands or simple questions, it might suggest that you prefer efficiency and quick results. You may have a busy lifestyle where you prioritize getting things done swiftly.

AI AND THE CONCEPT OF SELF

It's not just limited to machines doing tasks; it is involved in shaping how we see ourselves. Personal identity refers to our understanding of who we are as individuals.

A blog post that explores the effect of AI on identity acknowledges that this includes our beliefs, values, and the roles we play in society (Stevens, 2023). And, AI influences this understanding in various ways.

Identity in the Digital Age

One main way AI affects personal identity is through social media. Many platforms we use daily are powered by AI algorithms. These algorithms curate the content we see based on our previous interactions.

For example, if you often engage with fitness posts, the AI will show you more content related to fitness. This can lead to an altered self-image. Individuals might start defining themselves based on the content that AI promotes. They may feel pressure to conform to the ideals presented online, which can shift their personal identity.

Another example is the use of AI in personal assistants, like Siri or Google Assistant. These technologies understand our preferences and help us make decisions, from what music to listen to, to what to wear.

Over time, we rely more on these AI tools, which can impact our decision-making processes. Some may feel a sense of loss over their autonomy. They may question if their choices are truly theirs or influenced heavily by AI recommendations. This changing dynamic can reshape our personal identity and how we view our agency in the world.

AI and Its Impact on Social Identity

Social identity is about how we perceive ourselves in relation to groups and communities. AI plays a role in forming and shifting these identities through various channels.

One important channel is how we connect with others. Online platforms that utilize AI can help individuals find communities based on shared interests. For example, AI can recommend forums or groups where people discuss niche topics. This can be a good thing because it allows people to connect with like-minded individuals.

AI can also create echo chambers. An echo chamber is when people only hear opinions that are similar to their own, ignoring other perspectives. When AI shows us content based only on our beliefs, it can limit our understanding of the world.

This can affect how we see ourselves and strengthen stereotypes, making it harder to connect with different groups. People may start to view themselves only within a narrow circle of similar ideas and experiences, which can slow down personal growth and limit social understanding.

AI and Self-Perception

As we interact with AI more frequently, it starts to affect how we see ourselves. This influence can be subtle yet powerful. The way we perceive our abilities and worth can change based on these interactions.

The Dual Nature of AI Feedback

AI offers feedback that can enhance our skills. Consider a student using an AI-driven learning platform. When the student receives immediate responses on quizzes or exercises, they quickly learn from their mistakes.

This reliance can create a scenario where the person feels inadequate without the technology. It's essential to recognize that while AI can help build skills, overreliance may hinder self-confidence.

Celebrating Individual Growth

It's important to recognize personal growth without relying on AI. Noticing our own efforts can boost our self-esteem. People can keep a journal to track their improvement in different skills and note their personal successes, no matter how small.

Engaging in community groups for skill development can help people feel like they belong and find support. In these groups, members talk about their experiences and challenges, creating a teamwork atmosphere.

This group growth can lessen the need for technology to feel good about oneself. Having discussions can help people understand and recognize what they are capable of.

HUMAN–AI RELATIONSHIPS

The connection between humans and AI is complicated and always changing. People often form emotional bonds with AI, whether it's through personal assistants, chatbots, or smart systems used in different areas. This bond can come from how people communicate with AI.

Emotional Bonds With AI

Human attachment to AI occurs for several reasons. These reasons are rooted in the way AI interacts with us and the roles it plays in our lives. To truly understand this attachment, we must explore the various factors that contribute to it.

Understanding the Connection

Consider a conversational AI app like Replika. This app learns from our interactions over time. When we chat with Replika, it begins to understand our personality and preferences. It might learn that we enjoy discussing certain topics or that we often seek advice on specific issues.

This level of personalization makes users feel a deeper connection with Replika. They start to see it as more than just a chatbot; it becomes a companion that truly understands and supports them.

The Role of Convenience

Another important factor is convenience. AI systems can save us time and effort in our daily activities.

Here's a look at the factors that make our interactions with AI convenient:

- **Adaptability to mood:** Some AI can understand how users feel and respond in a way that fits their emotional state, showing empathy.
- **Always available support:** AI is always there to listen and help whenever someone needs to talk, making it a reliable source of support.
- **Anonymity:** Talking to AI allows people to share personal topics without worrying about their identity being revealed.

- **Availability of resources:** AI can provide helpful information and strategies for mental health, supporting emotional well-being.
- **Companionship:** AI provides a sense of connection for those who feel lonely, offering a reliable companion without the complications of human relationships.
- **Consistency in responses:** AI gives steady and predictable emotional responses, which can be comforting for people looking for reassurance.
- **Learning and growth:** Interacting with AI helps individuals learn about their feelings and thoughts in a safe space, supporting personal development.
- **Nonjudgmental interaction:** People can comfortably share their thoughts and feelings with AI, as there's no fear of being judged.
- **Reducing social anxiety:** For those who find social interactions difficult, talking to AI can make it easier to express themselves without pressure.
- **Personalized engagement:** AI can remember previous chats and preferences, making users feel recognized and connected.

Emotional Engagement

Emotional engagement also plays a critical role in human attachment to AI. Many AI companions, such as chatbots or robotic pets, are designed to evoke emotional responses. People may form bonds with these entities because they offer companionship or emotional support.

For instance, a study suggests that elderly individuals who interact with Pi.ai, an emotionally intelligent AI designed to provide companionship, report feeling significantly less lonely.

The AI, despite not being a living being, effectively offers a sense of comfort and connection, highlighting the human desire for social interaction, even when it is with a machine.

Ethical Implications

The concept of building a relationship with AI can appear distant or unfriendly to someone who is not used to it. For many, AI seems like a machine that lacks feelings and cannot genuinely connect with people.

This perspective comes from a lack of understanding about what AI truly is and how it functions in our lives. AI is not just a collection of complex algorithms; it is also designed to learn from interactions and improve over time, creating an evolving bond with users.

AI AND THE FUTURE OF HUMAN IDENTITY

The future concepts of identity are also influenced by AI. As technology advances, the lines between virtual identity and physical identity may blur. People may begin to adopt multiple identities across various online platforms.

For example, someone might be a professional on LinkedIn, a gamer on Discord, and a casual socialite on Instagram. Each of these identities may represent different facets of their personality, but they can also create confusion about who someone truly is.

Evolving Identities

This phenomenon reflects the evolution of identity in the digital age, where individuals curate multiple online personas that may not fully align with their real-life selves.

Here are discussions on future concepts of how AI is reshaping human identities:

Future Concept	Overview	Companies That Already Practice It
Multiverse theory	This theory suggests that many parallel universes exist, each representing different choices and outcomes.	Gaming companies like Insomniac Games (e.g., "Spider-Man: Miles Morales") explore multiverse concepts in storytelling.
Quantum consciousness	Some scientists believe our consciousness may arise from quantum processes in the brain.	Companies in quantum computing, such as IBM and D-Wave Systems, are exploring the implications of quantum mechanics.
Nonlocality and entanglement	In quantum physics, particles can affect each other instantly regardless of distance, influencing connectivity.	Tech companies involved in global communication technologies like Zoom or Slack promote interconnected identities.
Digital immortality and mind–body problem	The mind–body problem questions the relationship between consciousness and the physical body, possibly leading to digital immortality.	Neuralink, founded by Elon Musk, works on brain–machine interfaces that may integrate human cognition with digital technology.
Holographic principle	This principle posits that the universe may be a hologram, leading to shared and interconnected experiences.	Companies in VR, such as Oculus (owned by Facebook) and Microsoft (with HoloLens), create immersive experiences related to perception.

Blurring Boundaries

Blending human and machine identities is an important topic covering many ideas and effects. It looks at how people work with machines and how these interactions influence who we are. Unlike before, when machines were just tools to help us with everyday tasks, we now live in a time when machines can actually improve our abilities.

Take, for instance, the case of Cognixion. It focuses on developing technology that integrates neural signals and brain–computer interface (BCI) systems to assist individuals with disabilities in communicating and controlling devices.

Its goal is clear and important: to improve life for people with mental and physical challenges. Its products include Cognixion ONE, a wearable device that helps people communicate, and Environmental Control, a system that allows users to manage electronic devices and smart home systems using their thoughts.

EDUCATIONAL APPLICATIONS
OF AI

H ave you seen the movie *Limitless?*
It's a thought-provoking film that explores the concept of unlocking the full potential of the human brain. The story revolves around a struggling writer named Eddie Morra. He is portrayed by Bradley Cooper, who finds his life transformed when he comes across a mysterious pill. This pill allows him to access and use 100% of his brain's capacity.

When Eddie takes the pill, he experiences a dramatic change. He becomes incredibly intelligent and focused, which helps him write his book in just a few days.

The way I see it, a person's potential has no limits. Everyone has special skills and talents that can grow over time. As we improve our technology and understand ourselves better, we are just starting to discover what we can achieve. By focusing on personal growth and using AI wisely in our lives, we can explore many new opportunities ahead.

In this chapter, we'll now move on to the educational applications of AI. After we talked about the subject of AI and human identity, we're now ready to go further and explore more about what AI can do for education.

AI IN PERSONALIZED LEARNING

Each person learns differently. The way we understand information, process it, and retain it varies significantly from one individual to another. Some might prefer visual aids, while others thrive in auditory learning environments.

Various factors affect how we learn, and recognizing these differences can lead to more effective education and personal development strategies.

Adaptive Learning Systems

Recognizing that each person has a unique learning style is essential. Conducting assessments to determine individual learning preferences can lead to more tailored educational experiences.

For instance, a teacher might begin the year by asking students to complete a short questionnaire to identify their preferred learning style. Based on the results, they can make adjustments to their teaching strategies to reach every student effectively.

Different Types of Learners

There are many ways people learn new information, and understanding these different types of learners can help create a better learning environment. Each learner has unique needs and preferences. By recognizing these differences, teachers, parents, and students themselves can choose methods that work best for each individual.

Visual Learners

Visual learners are those who prefer to see information presented in a visual format. They might like to use diagrams, charts, or pictures to understand concepts better.

For example, when studying a topic like the solar system, a visual learner may benefit from looking at a colorful chart that shows the planets and their relationships to each other, rather than just reading about them in a textbook.

One effective way for visual learners to enhance their studies is through the use of mind maps. Mind maps allow learners to take notes visually by creating a diagram that connects related ideas.

This method can help them organize their thoughts and see how different ideas are connected. Flashcards with pictures or colors can also be helpful for these learners, as they can provide a visual cue that helps to trigger their memory.

Auditory Learners

Auditory learners learn best through listening. They may thrive in environments where they can discuss topics aloud or listen to lectures. For instance, instead of reading a book about history, an auditory learner might prefer to listen to a podcast or watch a documentary with narration.

One effective strategy for auditory learners is to engage in group discussions. Talking through ideas and hearing others' perspectives can reinforce their understanding.

Also, recording lectures or discussions to listen to later can be beneficial. They may find it helpful to read aloud or explain what they have learned to someone else, as this can help solidify the information in their mind.

Kinesthetic Learners

Kinesthetic learners, also known as tactile learners, are those who learn best through physical activity. They often benefit from hands-on experiences where they can manipulate materials or perform tasks.

For example, when learning about the human body, a kinesthetic learner might find it helpful to use models or participate in a lab activity.

To support kinesthetic learners, teachers can incorporate movement into their lessons. Simple activities like role-playing can help them understand complex concepts. In addition, using interactive tools, such as building blocks or simulation games, can make learning more engaging for these types of learners.

Reading/Writing Learners

Reading/writing learners prefer to engage with text. They find value in reading books, taking notes, and writing essays. These learners often excel when they can read extensively and write about what they have learned.

For instance, a reading/writing learner studying literature may enjoy reading different novels and then writing a summary of each one to reinforce their understanding.

To help these learners thrive, they should be encouraged to maintain a reading journal where they note down thoughts and reflections on their readings. Another effective strategy could be to assign written projects that encourage deep analysis and critical thinking.

This could include research papers or reflective essays that challenge their understanding. Providing a variety of reading materials can also stimulate their interest and improve their learning experience.

Social Learners

Social learners, or interpersonal learners, thrive in collaborative environments. They enjoy working with others and often learn through interaction and teamwork. For instance, group projects can significantly benefit these learners, as they can share ideas, ask questions, and gain insight from their peers.

Encouraging social learners to participate in study groups or peer teaching sessions can enhance their understanding of the material. These learners may also find it helpful to engage in discussions or group debates over the concepts being learned. Being in an environment where they can exchange opinions and perspectives can motivate them and make the learning process more enjoyable.

Solitary Learners

Solitary learners, or intrapersonal learners, find the best learning happens alone. They prefer to reflect internally and work independently. This type of learner often enjoys self-study, where they can dive deep into a topic without distractions.

To support solitary learners, it's essential to provide them with opportunities for independent study. Setting personal goals and allowing them time to explore concepts at their own pace can lead to a deeper understanding.

Journaling can also be an excellent tool for solitary learners, allowing them to express their thoughts and clarify their understanding of subjects. They may also appreciate online courses that allow for self-directed learning.

Combination Learners

You also need to recognize that some people aren't one learner type and don't fit into one category. They may exhibit a combination of learning styles, making them combination learners. A person might be predominantly a visual learner but also benefit from auditory methods or kinesthetic activities, depending on the subject matter.

To cater to combination learners, it is helpful to diversify teaching methods. A few sessions could incorporate visual aids, discussions, and hands-on activities related to a single topic.

This variety lets learners interact with the material in different ways, which helps them learn better. Being flexible in teaching methods can make the learning experience better for those who use different learning styles.

Benefits and Challenges

Using AI in education is helpful and takes learning up a notch.

Here are the benefits and challenges involved:

Benefits	Challenges
AI could analyze students' skills, interests, and job trends to give personalized advice on future careers.	Collecting personal data can raise issues about who has access and how it's used.
AI could create all sorts of social scenarios for students to practice and improve their emotional skills, helping them deal with people better in tomorrow's workplaces.	AI might struggle to capture the nuances of human interactions, affecting the quality of training.
AI could adapt lessons to reflect local cultures and histories, ensuring that the education is relevant to students' own lives.	Overemphasizing local content might limit students' exposure to broader ideas and worldviews.

INTELLIGENT TUTORING SYSTEMS

AI technology has transformed various fields, and one of its standout features is the ability to deliver personalized instructions and feedback. This means that, rather than offering a one-size-fits-all solution, AI can adapt to the unique needs of each user. This personalization makes learning and improvement much more effective.

For example, consider a student using an AI tutoring program like Khan Academy's Khanmigo. The system can analyze the student's past performance, identify areas where they struggle, and design lessons specifically for those weaknesses.

Creating a Safe Learning Environment

One often-overlooked benefit of utilizing intelligent tutoring systems is the safe learning environment it creates. Many users feel anxious about making mistakes, especially in traditional class-room settings. AI doesn't have the biases or expectations that human instructors might carry, and it allows users to fail without pressure.

For example, if a user is learning a new language, they may feel embarrassed when speaking in front of their peers. But with AI, they can practice out loud, receive corrections on pronunciation, and try again without fear of judgment. This promotes a culture of learning from errors rather than fearing them.

Tracking Progress Over Time

With AI's capability to monitor user performance, it can show progress over time. This tracking is vital for motivation. Users can see how much they have improved in specific areas, which encourages them to continue.

For instance, a musician using an AI practice tool might see a chart that outlines their progress in mastering a song. As they practice and receive actionable feedback, they can visually comprehend their journey and feel accomplished each step of the way.

Breaking Language Barriers

For users learning new languages, personalized AI can be invaluable. AI tools often provide not just vocabulary lists but also contextual examples and conversations suited to the learner's current skill level.

For instance, someone beginning to learn Spanish may receive exercises that focus on everyday phrases and essential vocabulary, while an intermediate learner might engage in dialogue simulations around different themes like travel or business. The AI continuously evolves the complexity of the sentences and vocabulary based on the learner's progress.

Design and Functionality

An intelligent tutoring system using GPT-4 can help a kid learn trigonometry in a friendly way. The catch is that it has to be designed well.

Here's how something like it works:

- **Personalized learning:** GPT-4 can find out what the student knows about trigonometry and what they need help with. It then adjusts its teaching to fit the student's needs.

- **Voice teaching:** The system can use voice to explain trigonometry concepts like sine, cosine, and tangent, making it like having a real tutor talking to the student.
- **Step-by-step help:** If the student has a problem, GPT-4 can walk them through the steps to solve it. For example, if they need to find the sine of an angle, it can explain how to use the unit circle to get the answer.
- **Practice problems and feedback:** The system can create practice problems for the student. After they try a problem, GPT-4 gives instant feedback, pointing out mistakes and how to fix them.
- **Visual aids:** Although it mainly uses text, GPT-4 can describe how to draw pictures or graphs to help the student see trigonometry functions more clearly.
- **Encouragement:** The system can cheer the students on to keep them motivated. It tracks their progress and celebrates their achievements.
- **Refinable:** The tutoring system is adjustable, meaning it can change based on how well the student is doing, making their learning experience better.
- **Resource suggestions:** GPT-4 can recommend videos, websites, or books that can help the student learn more about trigonometry.

PSYCHOLOGICAL EFFECTS OF AI IN EDUCATION

For individuals who have a strong desire to learn, AI can create a powerful psychological impact. This impact can manifest in various ways, affecting both motivation and engagement in learning processes. Understanding how AI influences learners can help us see its potential benefits.

Motivation and Engagement

When learners encounter AI tools, they often feel a boost in motivation. This increase in motivation can be attributed to the personalized learning experiences that AI provides.

For example, AI-powered educational platforms can adapt the content to match an individual's pace and skill level. A student struggling with math concepts can receive tailored exercises that target their specific challenges, encouraging them to overcome obstacles without feeling overwhelmed.

Engagement is another important aspect of the learning process that AI can significantly enhance. Many AI systems use gamification elements, turning learning into a fun and interactive experience. When tasks are rewarded with points, badges, or levels, students are more likely to participate actively.

For instance, language learning apps often incorporate games that challenge users to achieve certain goals. This method keeps learners interested and encourages them to practice more frequently, leading to improved skills and knowledge retention.

Teacher–Student Dynamic

With the introduction of AI into classrooms, this dynamic is evolving in remarkable ways. Teachers are finding that they can use AI to enhance their teaching methods, and students are experiencing new forms of learning that were not previously possible.

Below, let's explore this subject further:

Teacher's Role

Traditional Teacher Role	Teacher as a Facilitator
Delivers lessons primarily to students	Guides students in their learning journey
Focuses on lecturing and content delivery	Encourages discussion and critical thinking
Grades assignments and provides feedback	Uses AI to handle grading, giving more time for individual support
Less interaction with each student	More one-on-one interaction to meet individual needs
Students learn passively	Students engage actively and take charge of their own learning

Student's Role

Traditional Student Role	Student as an Active Learner
Receives information passively	Actively seeks out all sorts of information
Follows teacher's instructions without question	Explores topics and makes choices about learning
Relies on the teacher for help	Uses AI tools for assistance and problem-solving
Limited interaction with peers	Collaborates and engages with classmates in group work
Focused on grades and tests	Values learning and personal growth over just grades

FUTURE DIRECTIONS AND CHALLENGES

W e just capped off a chapter about the educational applications of AI. And, because you've heard my take on AI in education, let's now tackle the future directions and challenges that lie ahead.

As a means of beginning this chapter, let's talk about swarm intelligence. Swarm intelligence is a concept that comes from observing the behavior of social animals. When you look at insects like ants, bees, or termites, you can see how they work together in a group. They achieve tasks that seem impossible for an individual to manage alone.

For example, ant colonies can find the shortest path to food sources by using simple rules and communication methods. This form of collective behavior allows them to efficiently gather food and return to their nest.

One of the key aspects of swarm intelligence is that it is decentralized. This means no single leader is directing the group's actions. Instead, each individual makes simple decisions based on local information.

This decentralized approach helps the group adapt to changing conditions and make better decisions as a whole.

It's important to know this because it gives you a preview of the future. In the future, there are likely swarm intelligence algorithms, like particle swarm optimization and ant colony optimization. And, these can help you deal with the trickiest problems that need efficient solutions.

Futuristic problems require futuristic solutions, right?

THE FUTURE OF AI AND PSYCHOLOGY

Looking to the future, the collaboration between AI and psychology holds great potential. Enhanced products and services can emerge from this partnership, leading to better health outcomes, improved educational tools, and optimized workplace environments.

Interdisciplinary Research

One concept worth exploring in interdisciplinary research is neuroscanning. This focuses on capturing brainwave patterns during sleep with neuro-enhancement headsets, allowing for a direct connection to the subconscious. It sets an intriguing foundation for exploring dreams and opens up fascinating possibilities while remaining concise.

NeuroSky is a pioneering company that specializes in the development of BCI technologies. Their innovative products are all about electroencephalography (EEG) to monitor brain wave activity, translating mental states into actionable data.

With new developments in BCI technology, NeuroSky leads in helping people use their brain data in new ways. This technology could be useful for understanding dreams, improving learning, and managing emotions, making NeuroSky an interesting company to follow in the growing field of neurotechnology.

Emerging Technologies

Almost everywhere you turn today, you will notice the pervasive influence of new technology inspired by AI. The integration of AI into big data analysis is changing the way we view and understand information.

What Is Big Data?

Big data refers to vast amounts of structured and unstructured data that can be analyzed for insights and trends. This data comes from various sources, like social media, online transactions, sensor data, and more.

It's not just about having a large quantity of data; it's about how we process and analyze it to gain useful information. According to an article, it's because of this that there are key characteristics to acknowledge (Gillis & Robinson, 2021).

Here are the Five Vs of big data:

Characteristic	Description
Volume	The vast amount of data generated, ranging from terabytes to petabytes.
Velocity	This refers to how data is created and processed quickly, often right away.
Variety	These are the different formats of data, including structured, semi-structured, and unstructured data.
Veracity	The accuracy and reliability of the data, ensuring it can be trusted.
Value	The meaningful insights gained from analyzing data, informing decisions and practices.

Practical Applications of AI in Big Data

A great example of AI's impact on big data is in healthcare. Hospitals and clinics today generate massive amounts of data daily.

Using AI, healthcare providers can look at a lot of information to help patients get better care. AI can quickly find patients who might need help, which allows for faster treatment.

For instance, ML models can be trained to predict which patients are likely to develop complications from certain diseases based on their historical data. This predictive analytics approach not only saves lives but can also reduce healthcare costs by focusing resources where they are needed most.

Another area where big data and AI intertwine is public safety. Law enforcement agencies can analyze crime data to identify hotspots and predict when and where crimes might occur. By studying past crime patterns, AI can assist in deploying resources more effectively, thereby improving community safety.

EMERGING TRENDS

The future looks rather promising for AI and AI users, and this adage has been said too many times. But, just because it has been said a lot doesn't mean it's any less true.

XAI

One emerging trend is XAI. As mentioned in the introduction, it stands for explainable AI. It's a concept in AI that aims to make the operations of AI systems understandable to humans.

In simple terms, it is about creating AI models that not only make decisions but also provide clear reasoning behind those decisions. This is important because as AI continues to grow in influence and application across various fields, understanding how these systems arrive at their conclusions becomes crucial for trust and accountability.

What Is XAI?

XAI is about being open. An AI model can look at data, find patterns, and make choices. But when we can't see how it works—like a black box—we might not trust it.

For example, XAI helps people understand why an AI system denied a loan. Was it because of their credit score, financial situation, or bias in the system? This transparency makes AI systems more reliable.

Why Is XAI Important?

The significance of XAI stretches across various sectors, including healthcare, finance, law enforcement, and technology.

In affiliate marketing, for instance, marketers must disclose the data that informs their promotional strategies. If an affiliate program suggests a particular product to promote, marketers need to understand the reasons behind those recommendations to make informed choices.

Similarly, when analytics tools highlight trends in consumer behavior, it is essential to comprehend the factors leading to those insights. Providing transparency builds trust among consumers and partners, which is vital for the successful integration of affiliate marketing practices.

AI and Neuroscience

With the help of AI, researchers can analyze large amounts of data much more quickly and effectively than before. This is important because understanding how the brain works is a complex task that involves many variables.

For instance, AI algorithms can process brain scans and patterns of brain activity to highlight areas that are involved in specific functions. This can lead to better insights into everything from memory to mood regulation.

The Basics of Brain Function

To grasp how AI enhances our understanding of the brain, it's essential to know a bit about the brain itself. The brain is made up of billions of cells called neurons that communicate through connections known as synapses. Each neuron can connect to thousands of other neurons, forming networks that are responsible for our thoughts, emotions, and actions. It can be overwhelming to think about all these interactions. Here, AI tools can break things down into more manageable parts.

How AI Works in Neuroscience

AI helps scientists to sift through the massive amount of data produced by brain research. Traditional analysis methods can be slow and often miss crucial patterns. ML can identify patterns in data that might not be obvious to the human eye.

For example, researchers can upload thousands of images from brain scans to an AI system. This system can learn to recognize normal brain activity versus abnormal activity, which can help in diagnosing conditions like Alzheimer's disease or schizophrenia.

Examples of AI Applications in Neuroscience

One specific example of AI's application in understanding brain function is the use of it in studying brain tumors. When doctors detect a tumor, they can use AI to analyze images of the tumor and the surrounding brain tissue. The AI can help predict how the tumor will grow and respond to treatment.

PSYCHOLOGICAL CHALLENGES IN AI ADOPTION

Despite the clear potential benefits of AI, resistance to change can hinder adoption.

Resistance to Change

Employees, for one, may feel threatened by new technologies or worry that AI will replace their jobs. To combat this resistance, organizations should focus on communication and transparency.

Sharing success stories can also help reduce fears about using AI. Showing examples of other companies that have successfully used AI can inspire hesitant employees. Seeing real benefits, like faster work or better choices, can encourage staff to accept AI technology.

Finding the Right Solutions

Identifying the right AI solutions is vital for adoption. With so many options available, businesses need to make informed choices that align with their specific needs.

Researching different types of AI tools, such as ML models, NLP systems, or robotic process automation, can help companies determine what might work best for them.

AI and Human Autonomy

Human autonomy refers to the ability of individuals to make their own choices and decisions without undue influence or control from external forces. When we introduce AI into our lives and workplaces, we must ensure that it enhances human autonomy rather than diminishes it.

Combining Strengths

One effective way to strike a balance between AI and human autonomy is by leveraging the strengths of both. Organizations can encourage a partnership approach where AI handles data-driven tasks while humans focus on interpersonal connections and emotional intelligence.

For example, in customer service, AI-driven chatbots can manage routine inquiries, allowing human representatives to handle more complex situations that require empathy and understanding. This collaboration not only improves service quality but also retains the essential human touch that customers value.

Engaging in Dialogue

Creating an ongoing dialogue about the use of AI in society will help maintain awareness of its impact on human autonomy.

Conversations involving policymakers, technologists, ethicists, and the public can foster a deeper understanding of the implications and challenges of integrating AI into our lives.

Open forums can serve as platforms to discuss balancing AI's capabilities with the need to maintain human agency. Engaging diverse voices in these discussions can help reflect varied perspectives and ensure that multiple viewpoints are considered.

CASE STUDIES AND REAL-WORLD APPLICATIONS

I n this final chapter, let's talk about the case studies and real-world applications of AI.

And, speaking of case studies, one striking case study revolves around training AI and accomplishing this by playing a series of chess games.

Leela Chess Zero is an AI that learns to play chess by playing many games. Just like a human gets better with practice, Leela gets better by looking at past games and changing its strategy. It uses reinforcement learning, which means it plays against itself to improve its understanding of chess.

The more it plays, the better it gets, learning complex tactics that even experienced players might take years to learn. This example shows how AI can improve by playing games and highlights its ability to learn from experience.

CASE STUDY 1: AI IN HEALTHCARE

AI can help quickly and accurately analyze and interpret the large amounts of data produced in medical settings.

AI in Diagnostics and Treatment

For instance, AI algorithms can look at medical images, such as X-rays or MRIs, to identify abnormalities. This process is often faster than traditional methods, which can help in providing quicker diagnoses for patients.

One concrete example is the use of AI in radiology. An AI system can be trained with thousands of X-ray images to recognize patterns associated with conditions like pneumonia or fractures. When a new X-ray image is introduced, the AI can compare it against the patterns it has learned.

If the AI detects something unusual, it can alert radiologists, who can then confirm the findings through further examination. This not only speeds up the diagnosis process but can also reduce errors that may occur with human interpretation alone.

AI in Treatment Recommendation

Once a disease is diagnosed, the next important step is to determine the best treatment plan. Here again, AI plays a significant role in recommending personalized treatment options.

These AI systems analyze the patient's medical history, genetic information, and other relevant data, making them ideal for identifying treatment strategies that have the highest probability of success.

For example, in oncology, AI tools can analyze genetic data from tumors. By looking at specific mutations, the AI can suggest therapies that target those mutations.

This helps doctors choose the most effective drugs for their patients rather than relying solely on generalized treatment protocols. Personalized medicine is becoming more prevalent, and AI is at the forefront of this transformation.

AI's Impact on Efficiency and Cost Reduction

The incorporation of AI in diagnosing diseases and recommending treatments also brings significant benefits in terms of efficiency and cost. By speeding up the diagnosis process and reducing unnecessary procedures or tests, healthcare providers can save time and resources. For example, AI can help triage patients in emergency rooms by quickly assessing which cases require immediate attention.

In addition, the reduction of diagnostic errors minimizes costs associated with misdiagnosis. When AI assists in detecting diseases accurately, it can prevent wasted spending on ineffective treatments and help allocate resources more effectively. This means less financial burden on both healthcare facilities and patients.

Ethical Considerations and Transparent AI

While the advancements in AI for healthcare are promising, it is essential to highlight the ethical considerations involved. There is a growing concern about how these AI systems make their decisions. There needs to be transparency in how AI diagnostics are derived so that healthcare professionals understand and trust the recommendations made by these systems.

Psychological Impacts on Patients and Healthcare Providers

Patient data privacy needs to be a top concern. AI systems need access to important medical information to work well. Healthcare providers have to find a way to protect patient data while still being able to analyze it effectively.

CASE STUDY 2: AI IN BUSINESS AND MARKETING

Understanding customer behavior is a key factor in the success of any business. It involves studying how customers make decisions about what to buy and why they choose certain products or services over others.

For instance, when a business recognizes that its customers value sustainability, it can focus on promoting eco-friendly products. This tailored approach not only attracts the right customers but also builds trust and loyalty.

Customer Insights and Personalization

Customer insights and personalization in marketing are about making customers feel valued and understood. When customers receive personalized messages or offers, they are more likely to engage with a brand.

For example, an online retail store might send personalized product recommendations based on a customer's previous purchases. This simple act can significantly increase the chances of a sale.

Creating Personalized Marketing Strategies

After gathering and analyzing data, create personalized marketing strategies. One effective approach is segmentation. This involves dividing customers into groups based on shared characteristics, such as age, location, or interests.

For example, a clothing retailer might segment customers into categories like "young adults" and "families." By doing this, the retailer can tailor its marketing messages to each group. Young adults might appreciate trendy styles and social media promotions, while families might prefer discount offers on bulk purchases.

Another strategy is to use retargeting ads. Retargeting allows businesses to display ads to customers who have previously visited their website but did not make a purchase.

For example, if a customer looked at a dress or bike but left the site without buying, the retailer can show ads for that dress or bike as the customer browses other sites. This method keeps the product fresh in the customer's mind and encourages them to return and complete their purchase.

Measuring Success in Personalization

To determine if personalization efforts are successful, businesses need to measure the right metrics. This could include tracking open rates on email campaigns, click-through rates for personalized ads, and conversion rates for specific offers. These metrics provide measurable insights into how effective the personalization strategies are working.

For instance, if a company notices that personalized email campaigns have higher open rates than generic ones, it's a clear indicator that customers respond better to personalized content.

Another example is analyzing the impact of targeted ads; if a retargeting campaign leads to a significant increase in sales, it confirms that reminding customers about a product is effective.

Adapting to Changes in Customer Behavior

Customer behavior is not static; it changes over time due to various factors like market trends, economic conditions, or shifts in societal values. Businesses must remain flexible to adapt their marketing strategies accordingly. This means continuously monitoring trends and being willing to change tactics.

For example, during the COVID-19 lockdowns, many consumers shifted to online shopping. Companies that quickly adapted to this change by improving their online presence and offering convenient delivery options thrived.

Those that did not may have struggled to retain customers. Thus, staying informed about changes in customer behavior is essential for long-term success.

Employee Management and Automation

Managing workforce efficiency is an almost incomparable aspect of running a successful organization. It refers to how well a company's employees perform their tasks and responsibilities in relation to the resources available.

High efficiency means that employees use their time and skills productively, which ultimately leads to achieving the company's goals. To understand workforce efficiency better, it's important to break it down into manageable parts and consider various strategies that can help enhance it.

Understanding Workforce Efficiency

You can define workforce efficiency as the ratio of productive output to the total resources used. It looks at how effectively human resources are utilized. For example, if an employee spends a lot of time on a task but does not produce quality results, the efficiency is low.

Conversely, if an employee can complete their tasks quickly and effectively, the efficiency is high. Managers need to find a balance where employees are not overworked, yet can perform their jobs effectively.

Encouraging Open Communication

Open communication is important for a good work environment. Employees should feel safe to share their thoughts, problems, and ideas. Regular meetings or feedback sessions can help make this communication easier.

For instance, using a suggestion box can allow employees to voice their concerns anonymously. This feedback can help management identify problem areas and find solutions to enhance efficiency.

Creating a Supportive Work Environment

A supportive work environment enhances employee performance. According to Wingmore (n.d.), 70% of employees report increased productivity when they feel cared for, appreciated, and supported by the people they work for.

In business, you can refer to this as the 70% rule of productivity. If you look at matters from the side of upper management, this is a practical move. If you look at matters from the side of the employees, it's also equally practical because it encourages them to be the best versions of themselves.

For example, team-building exercises can foster relationships among team members. An environment where employees support each other can lead to increased productivity as they work together toward common goals.

Prioritizing Employee Well-being

Prioritizing employee well-being is also essential for maintaining workforce efficiency. Employees need breaks, too, and those who always feel well physically and mentally are more productive. Encouraging time-outs and offering programs for mental health can help employees manage stress.

For instance, introducing flexible work hours or remote working options can allow employees to create a schedule that works best for them. When employees feel their well-being is valued, they are likely to be more committed and efficient in their roles.

CASE STUDY 3: AI IN SMART HOMES AND CITIES

Home automation and security have seen significant improvements due to the advancement of AI. Many homeowners now find themselves using these technologies to make their lives easier and safer.

Smart Home Technologies

The use of AI in this field is not just a trend; it's becoming a standard aspect of modern living. Through AI applications, homeowners can manage various aspects of their homes, from lighting to security systems, with minimal effort.

Smart Lighting Control

One of the most common applications of AI in home automation is smart lighting. These systems allow users to control the lighting in their homes remotely or through voice commands. Imagine coming home after a long day at work and finding your home well-lit without having to switch on the lights manually.

For instance, you can install smart bulbs that adjust their brightness based on the time of day. In the morning, they can gradually brighten to simulate a natural sunrise, helping you wake up more gently. At night, you can set them to dim gradually as you prepare for bed. This not only creates a comfortable atmosphere but also contributes to energy savings.

Enhanced Security Systems

Another area where AI is making a significant impact is in home security. Traditional security systems often rely on motion sensors and alarms. However, AI-powered security cameras can learn the patterns of normal activity around your home. For example, if a neighbor walks by your house every evening, the system can recognize this pattern and avoid triggering an alarm.

Furthermore, these smart cameras can send real-time alerts to your phone when they detect unusual activity. Suppose a stranger approaches your front door during the night; the system will notify you immediately. This ability to differentiate between usual and unusual activity not only reduces false alarms but also enhances your peace of mind.

Voice Assistants and Home Management

Voice assistants are also pivotal in home automation. Devices like Amazon Alexa or Google Assistant allow you to perform various tasks just by speaking. You can command these devices to turn on the coffee maker, adjust the thermostat, or lock the front door.

For instance, if you're entertaining guests and need to play a specific playlist, you can simply ask your voice assistant to do it for you. The integration of voice control with various smart devices means that your home becomes more interactive and responsive to your needs.

Energy Management

AI applications also contribute significantly to energy management within homes. Smart thermostats can learn your heating and cooling preferences and adjust themselves automatically.

For instance, if you usually lower the temperature at night, the thermostat can learn this habit and make the adjustments for you. This not only improves comfort but can also lead to lower energy bills.

Smart Appliances

The rise of smart appliances has transformed how we approach everyday tasks. From refrigerators that can track food inventory to washing machines that can be scheduled for optimal times, these devices use AI to streamline household chores.

AI-enabled washing machines can learn your laundry habits, recommending the best wash cycle based on the load size and type of fabric. Many of these appliances can be controlled remotely, allowing you to start a load of laundry while you're at work or adjust settings to fit your schedule.

Home Monitoring Systems

Beyond traditional security, AI-powered home monitoring systems bring additional layers of safety and convenience. For example, some systems offer features like smoke detection and water leak alerts. If a detector senses smoke in your home, it can notify you right away and, in some cases, even contact emergency services if necessary.

Water leak sensors can alert you to potential problems, such as a leak in the basement before it causes significant damage. These alerts can be sent directly to your phone, so you can take action even if you are not at home.

Personalized Experiences

AI can also personalize your home automation experience. Smart systems can analyze your habits and preferences, suggesting adjustments to create a more comfortable living environment.

For example, if you typically enjoy watching TV in the evenings, the system could automatically adjust the lighting and temperature to your preferred settings as the evening approaches.

Smart City Initiatives

AI-driven urban environments can have a significant impact on the psychology of individuals living in these areas. An example is Singapore with its Smart Nation initiative, which integrates AI across multiple sectors to improve city living.

The city's safety improvements are impressive. It uses smart cameras that can spot unusual behavior and notify the police automatically.

Psychological Implications of AI-Driven Urban Environments

Moreover, the presence of AI can lead to changes in social interactions. In a city where AI is employed to facilitate communication and services, people might rely more on technology than on face-to-face interactions.

For instance, if there are mobile apps that help you find friends in crowded places, some may prefer to use these tools rather than navigate social situations independently. This reliance on technology can create a sense of isolation if people spend more time interacting through devices rather than with one another directly.

Social Implications of AI-Driven Urban Environments

The impact of AI on cities can be significant. Cities that use AI to improve safety might set up systems that watch over public areas and analyze data to help stop crime before it happens.

While this could improve safety, it also raises privacy concerns. Residents might feel they are constantly being watched, which can lead to anxiety and distrust among community members.

For example, a neighborhood that previously had a close-knit community might begin to feel divided if residents feel uncomfortable with the use of surveillance technologies.

HELP ME SPREAD THE WORD!

No matter how you use AI in your life right now or how you may come to work with it in the future, understanding the relationship between AI and psychology is going to be integral. Take a moment now to help even more readers reach that understanding.

Simply by sharing your honest opinion of this book and a little about what you've learned here, you'll provide a signpost that will help new readers find their way to it.

JUST ONE CLICK!

Thank you so much for your support. I hope you feel empowered to use AI to enhance your life: There's an exciting future ahead, and you want to be part of it!

Scan the QR code below

CONCLUSION

Now that we've reached the closing of this book, I hope you learned enough about using AI to your advantage. And, now that we're about to part ways, let's circle back to what I mentioned in the introductory chapter about the principle of reciprocity.

I want you to reflect on it as a reminder that it's a give-and-take approach. Remember, **you have to give** (i.e., you also need to maximize your potential and use it to guide the AI system) **before you can take** (i.e., receive a satisfactory outcome and maximize AI's potential).

Another important thing to remember is to be mindful of how you prompt or provide instructions. As the title of this book suggests, psychology plays a part in all this. While it may seem as if AI will be fine regardless of what you input, your approach actually matters a lot.

Did you know that a simple "please" and "thank you" can influence the quality of its output (for the better)?

To help me convey my point, consider putting yourself in AI's shoes. Would you perform to the best of your abilities if someone just barked orders at you? Or, would you prefer being gently given orders?

Chances are, it's the latter.

SUMMARIZING KEY TAKEAWAYS

Now, let's summarize what we've learned throughout. Remember, psychology offers a unique lens to explore how AI systems operate and their impact on human behavior and decision-making.

Integration of AI and Psychology

By analyzing the relationship between AI and psychology, we can learn how AI technologies can be designed. The objective is to align them more closely with human thought processes.

This understanding is straightforward. And, it can create AI systems that are more intuitive and user-friendly.

Implications for Practice

AI developers also benefit from the practical applications of AI. For one, they use ML techniques to create smarter and more efficient algorithms. They want algorithms that can help them carry out work in a more efficient manner.

One typical example is the development of recommendation systems. Such systems work like those used by streaming services or online retailers. These systems analyze user behavior to suggest relevant content or products, enhancing UX.

LOOKING AHEAD

The intersection of AI and psychology is a fascinating area of exploration. Call me a bit biased, but the way I see it, you're only scratching the surface if you know about psychology's relationship with AI. As technology continues to grow, it will provide us with innovative ways to understand and improve almost everything we do in life.

Future Trends in AI and Psychology

By focusing on ethics, working together across all sorts of fields, and using technology wisely, we can build a future where psychological help is within reach. It's also better and more available to everyone. This change will help us understand ourselves more and find new ways to promote mental well-being in our society.

Staying Informed

More and more people say that the future is AI, and I couldn't help but agree. For this reason, it will work to your advantage if you're always informed.

One effective way to stay informed about the many advancements in AI is through webinars. Webinars are online seminars that offer a platform where experts can share their knowledge and experience.

What Are Webinars?

Webinars can range from short presentations to longer, more in-depth sessions. During a typical webinar, participants can learn about current trends, research findings, and case studies related to AI.

Unlike traditional lectures, webinars often encourage participation from the audience. This interaction can happen through live chat or Q&A sessions, which allows attendees to seek clarification on topics they find challenging.

FINAL THOUGHTS

Building fair AI systems isn't simple. It involves taking deliberate actions throughout the development process.

One practical approach is to form diverse teams composed of individuals from various backgrounds and let those people be in charge of AI training. This diversity can provide different perspectives that help identify and mitigate potential biases in AI algorithms.

For instance, if a team only consists of individuals from similar backgrounds, they may unintentionally overlook the biases present in their AI model. To combat this, organizations can implement regular training programs focused on recognizing and addressing bias in technology.

Commitment to Ethical AI

Commitment to ethical AI also means staying in line and regularly checking AI systems. These checks can look at the data used to train AI models. For example, using different sources of data can help make sure the information is complete and shows various experiences.

APPENDICES

Throughout this book, you'll encounter terms and subjects you may want to learn more about. So, here's an extra chapter that I've dedicated to elaborate on the relevant information.

KEY AI AND PSYCHOLOGY TERMS

If you want to learn more about AI and psychology and are planning to read relevant books in the future, it's best to be familiar with some key terms.

Glossary of Terms

The purpose of compiling this glossary is for you to have a brush-up, especially in cases where you need clarification.

Here are the terms you want to check out:

- **Algorithm choice:** Selecting the most appropriate computational method for a specific task or problem.

- **Attention mechanisms:** Techniques allowing neural networks to focus on relevant parts of input data.
- **Autonomous systems:** Self-governing machines capable of independent decision-making and action.
- **Backpropagation algorithm:** A method for training neural networks by adjusting weights based on error gradients.
- **Bayesian decision theory:** A statistical approach to decision-making under uncertainty using probability theory.
- **Bias:** Systematic deviation from true values in data, models, or human judgment.
- **Big data:** This refers to vast amounts of structured and unstructured data that can be analyzed for insights and trends.
- **Cognitive load:** Mental effort required to process information and perform tasks.
- **Cognitive load theory:** Framework explaining how mental resources are used during learning and problem-solving.
- **Cognitive psychology:** Study of mental processes such as perception, memory, and reasoning.
- **Deep learning:** Machine learning approach using multilayered neural networks to learn hierarchical representations.
- **Diffusion models:** Generative AI models that create data by gradually denoising random noise.
- **Emotional recognition:** Ability to identify and interpret human emotions from various inputs.
- **Ethical AI:** Development and application of AI systems adhering to moral principles and societal values.
- **Explainable AI (XAI):** A way for humans to understand and trust machine learning algorithms. In other words, it's

about making AI models more transparent and understandable.

- **Game theory:** This is a mathematical study of strategic decision-making in competitive situations.
- **Information processing theory:** Cognitive model comparing human thought to computer data processing.
- **IQ tests:** This is a standardized assessment of tests designed to measure human intelligence and cognitive abilities.
- **Language understanding:** This refers to an AI system's capability to comprehend and interpret human language in context.
- **Misinformation:** False or inaccurate information spread intentionally or unintentionally.
- **Model development:** Process of creating, training, and refining machine learning models.
- **Multimodal models:** AI systems capable of processing and integrating multiple types of input data.
- **Narrow AI:** This is an AI system specialized for a specific task or domain.
- **Neural networks:** Computing systems inspired by biological brain structures, used in machine learning.
- **Neural Turing machines:** Computational models combining neural networks with external memory.
- **Path planning:** This refers to the process of finding optimal routes for autonomous agents in various environments.
- **Performance assessment:** This is an evaluation of how well a system, model, or individual accomplishes specified tasks.
- **Probabilistic graphical models:** Statistical tools representing complex relationships between variables using graphs.

- **Problem-solving:** Cognitive process of finding solutions to challenges or obstacles.
- **Robotics:** Interdisciplinary field focusing on design, construction, and use of autonomous machines.
- **Social-emotional learning (SEL):** Social-emotional learning is an educational approach that teaches students to recognize and manage emotions, develop empathy, build positive relationships, and make responsible decisions.
- **Spectral analysis:** Technique for decomposing complex signals into component frequencies.
- **Superintelligent AI:** This is a hypothetical AI surpassing human cognitive abilities.
- **Supervised learning:** This is an ML approach using labeled training data to make predictions or decisions.
- **Synaptic plasticity:** Ability of neural connections to strengthen or weaken based on activity and experience.
- **Swarm intelligence:** Collective behavior of decentralized, self-organized systems, often inspired by nature.
- **Tokenization:** This is a process of breaking text into smaller units (tokens) for natural language processing.
- **Transfer learning:** This is an ML technique about applying knowledge from one task to improve performance on another.
- **Transformer models:** Neural network architecture using self-attention mechanisms for sequence processing.
- **Unsupervised learning:** Machine learning approach finding patterns in unlabeled data without predefined outcomes.
- **Virtual reality (VR) simulation:** A computer-generated simulation of a three-dimensional environment that can be interacted with in a seemingly real way using special electronic equipment.

- **Working memory capacity:** Amount of information an individual can temporarily hold and manipulate.

Acronyms

Anywhere AI is concerned comes three of the most common terms: artificial intelligence (AI), machine learning (ML), and natural language processing (NLP). and, beyond these three are other common abbreviations and acronyms.

Here are some acronyms worth checking out:

- **Artificial general intelligence (AGI):** Hypothetical AI system capable of performing any intellectual task that a human can.
- **Cognitive behavioral therapy (CBT):** It's a mental health treatment that helps patients identify and modify negative thought patterns and behaviors to improve their emotional well-being and coping skills.
- **Convolutional neural networks (CNNs):** Deep learning architecture specialized for processing grid-like data, particularly effective in image analysis.
- **Differentiable neural computers (DNCs):** Neural network models that combine deep learning with external memory systems for complex reasoning tasks.
- **Long short-term memory (LSTM):** Recurrent neural network architecture designed to learn long-term dependencies in sequential data.
- **Monte Carlo tree search (MCTS):** Decision-making algorithm using random sampling to evaluate moves in complex scenarios, often used when playing games with AI.

- **Simultaneous localization and mapping (SLAM):**
Computational problem of constructing a map of an
unknown environment while tracking an agent's location
within it.

RECOMMENDED READING AND RESOURCES

Reading is an active way to learn that helps us think more critically and analyze information. Books and articles are usually
reviewed and edited carefully, especially in schools and universities. This means they are more reliable than other, less formal
learning sources.

Books and Articles

Books and articles are great ways to learn. They are useful tools
because they give detailed and organized information on many
topics. Readers can take their time with the material, which helps
them understand and think about it more deeply.

Here's a list of books and articles worth checking out:

Book	Strengths	Weaknesses
Mind and Machine: What It Means to Be Human by Michael L. Anderson	In-depth exploration of human cognition in contrast with machines	May be dense for general readers
The Singularity Is Near by Ray Kurzweil	Visionary insights into future tech and AI	Some concepts may seem overly optimistic
How to Create a Mind by Ray Kurzweil	Discusses the potential of AI and innovation in replicating human thought	Concepts can be speculative
Superintelligence: Paths, Dangers, Strategies by Nick Bostrom	Offers a critical view of AI development and its implications	Some argue it's alarmist

The Age of Spiritual Machines by Ray Kurzweil	Integrates technology with human experiences and spirituality	Can be challenging to digest
Homo Deus: A Brief History of Tomorrow by Yuval Noah Harari	Engages with future possibilities of humanity and technology	Broad scope may lack depth in specific areas
Weapons of Math Destruction by Cathy O'Neil	Analyzes the negative effects of algorithms on society	May feel overly focused on criticisms
Artificial Intelligence: A Guide to Intelligent Systems by Michael Negnevitsky	Practical insights into AI applications across industries	Technical focus may not appeal to all
The Three-Body Problem by Liu Cixin	Discusses philosophical and ethical implications of first contact with alien intelligence	The complex narrative may be challenging
Nanotechnology: Understanding Small Systems by Ben Rogers	A comprehensive introduction to nanotech and its potential	May be too technical for casual readers
The Master Algorithm by Pedro Domingos	Explores ML algorithms in-depth	Can be heavy on technical jargon
The Emotion Machine by Marvin Minsky	Explores emotions in machines and implications for AI	The theoretical framework may seem abstract
The Public Policy of Nanotechnology: Technologies in the Shadow of the Future by Michèle L. M. V. D. Beek	Discusses societal impacts of nanotechnology	Limited scope on technical details
Wired for War by P. W. Singer	Looks at the implications of robotics in warfare	May not appeal to those uninterested in military tech
Machines of Loving Grace by John Markoff	Investigates the relationship between humans and machines over time	May be too historical for some readers
The Fourth Industrial Revolution by Klaus Schwab	Explains the changes in society due to technological advancements	May feel too theoretical for practical applications
What Technology Wants by Kevin Kelly	Philosophical exploration of technology as an evolving force	Concepts can be abstract and philosophical

Reinventing Discovery: The New Era of Networked Science by Michael Nielsen	Discusses the impact of technology on scientific discovery	May seem niche to nonscientists

Online Courses and Workshops

Apart from reading books and articles, an effective way to deepen your understanding and skills in relation to the psychology of the mind and machine is to enroll in different online courses and workshops.

For this section, rather than provide you with the actual names of courses, I believe it's a much better idea to give you tips on how you can go out there and seek out courses. I'm going this route because there are so many great courses out there.

Here are ways on how you can seek courses and workshops:

- checking websites of universities known for AI and psychology research
- exploring AI-focused educational sites like Fast.ai or DeepLearning.AI
- looking at professional organizations like the American Psychological Association for workshops
- searching major MOOC platforms like Coursera, edX, and Udemy

LIST OF AI AND PSYCHOLOGY JOURNALS AND CONFERENCES

Apart from books, articles, online courses, and workshops, scientific journals and conferences also provide access to the latest research and advancements in various fields, ensuring readers are informed about developments and breakthroughs.

Journals

An advantage of AI and psychology journals is that they are peer-reviewed. This means that they're thoroughly checked to ensure the information on them is credible and reliable.

Here are journals worth checking out:

- *Journal of Artificial Intelligence Research (JAIR)*
- *Artificial Intelligence Journal (AIJ)*
- *International Journal of Machine Learning Research*
- *Neural Networks*
- *Journal of Machine Learning Research*
- *Nature Machine Intelligence*
- *AI Magazine*
- *Cognitive Computation*

Conferences

There are major conferences where AI and psychology meet. One benefit is the great chances to network. In-person conferences allow you to connect with other professionals, experts, and leaders in the field.

Here are conferences worth attending:

- International Conference on Machine Learning
- International Conference on Learning Representations
- Conference on Neural Information Processing Systems
- Association for the Advancement of Artificial Intelligence

REFERENCES

About. (n.d.). Twitch. https://www.twitch.tv/p/en/about/

Almakhour, M., & Mellouk, A. (2020). *Theorem prover.* ScienceDirect. https://www.sciencedirect.com/topics/computer-science/theorem-prover

Al-Masri, A. (2024, March 7). *Backpropagation in a neural network: Explained.* Built In. https://builtin.com/machine-learning/backpropagation-neural-network

Bickersteth, D. (2023, September 12). *Evolving identities: A holistic guide for individuals and organizations.* CARVE your Niche. https://medium.com/carve-your-niche/evolving-identities-a-holistic-guide-for-individuals-and-organizations-baef75e9b15

Brower, T. (2022, November 6). 70% aren't prepared for the future of work: Demands for upskilling surge. *Forbes.* https://www.forbes.com/sites/tracybrower/2022/11/06/70-arent-prepared-for-the-future-of-work-demands-for-upskilling-surge/

Brown, D. (2022, February 17). *AI market in healthcare expected to surpass $34B by 2025.* Health Exec. https://healthexec.com/topics/artificial-intelligence/health care-ai-market-surpass-34b-2025

Celemin, C., Pérez-Dattari, R., Chisari, E., Franzese, G., de Souza Rosa, L., Prakash, R., Ajanović, Z., Ferraz, M., Valada, A., & Kober, J. (2022, October 31). Interactive imitation learning in robotics: A survey. *ArXiv: 2211.00600* https://doi.org/10.48550/arXiv.2211.00600

Digital immortality and the future of consciousness: A deep dive into the concept of mind uploading. (2023, June 25). *The AI Blog.* https://www.theaiblog.net/digital-immortality-and-the-future-of-consciousness-a-deep-dive-into-the-concept-of-mind-uploading/

eLearning Company Blog. (2024, March 20). Putting ChatGPT to work: Enhancing student-teacher interactions in online learning. *eLearning Company.* https://elearning.company/blog/putting-chatgpt-to-work-enhancing-student-teacher-interactions-in-online-learning/

Flavin, B. (2019, May 6). *Different types of learners: What college students should know.* Rasmussen University. https://www.rasmussen.edu/student-experience/college-life/most-common-types-of-learners/

Forrester. (2021, September 7). Ginger and Headspace merge: Looks to build mental resilience in all employees by giving them more choice. *Forbes.* https://www.forbes.com/sites/forrester/2021/09/07/ginger-and-headspace-merge-looks-to-build-mental-resilience-in-all-employees-by-giving-them-more-choice/

Gandhi, T. K., Classen, D. C., Sinsky, C. A., Rhew, D. C., Vande Garde, N., Roberts, A., & Federico, F. (2023). How can artificial intelligence decrease cognitive and work burden for front line practitioners? *JAMIA Open, 6*(3). https://doi.org/10.1093/jamiaopen/ooad079

Gibbons, S., Mugunthan, T., & Nielsen, J. (2023, October 20). *The 4 degrees of anthropomorphism of generative AI*. Nielsen Norman Group. https://www.nngroup.com/articles/anthropomorphism/

Gillis, A., & Robinson, S. (2021, March). *The 5 Vs of big data*. TechTarget. https://www.techtarget.com/searchdatamanagement/definition/5-Vs-of-big-data

Ginger's mental health app is now Headspace Care. (n.d.). Headspace. https://organizations.headspace.com/ginger-is-now-part-of-headspace

Haan, K. (2023, July 20). Over 75% of consumers are concerned about misinformation from artificial intelligence. *Forbes Advisor*. https://www.forbes.com/advisor/business/artificial-intelligence-consumer-sentiment/

Haan, K. (2024, June 15). 24 top AI statistics and trends in 2024. *Forbes Advisor*. https://www.forbes.com/advisor/business/ai-statistics/

Hosna, A., Merry, E., Gyalmo, J., Alom, Z., Aung, Z., & Azim, M. A. (2022). Transfer learning: A friendly introduction. *Journal of Big Data, 9*(1). https://doi.org/10.1186/s40537-022-00652-w

The Investopedia Team. (2022, January 1). *Weak AI*. Investopedia. https://www.investopedia.com/terms/w/weak-ai.asp

Jennings, K., & Knapp, A. (2023, August 9). InnovationRx: 70% of U.S. adults "concerned" about AI in healthcare. *Forbes*. https://www.forbes.com/sites/alexknapp/2023/08/09/innovationrx-70-of-us-adults-concerned-about-ai-in-healthcare/

Kalpokas, I. (2023). Work of art in the age of its AI reproduction. *Sage Journals*. https://doi.org/10.1177/01914537231184490

Karl, T. (2024, February 12). DBSCAN vs. k-means: A guide in Python. *New Horizons*. https://www.newhorizons.com/resources/blog/dbscan-vs-kmeans-a-guide-in-python

Kowalski, P. (2021, March). *Why should 73% of Polish websites have a closer look at their mobile user experience?* Think with Google. https://www.thinkwithgoogle.com/intl/en-emea/marketing-strategies/app-and-mobile/why-should-73-of-polish-websites-have-a-closer-look-at-their-mobile-user-experience/

Lim, E., & Conversation, T. (2023, March 23). *The multiverse: How we're tackling the challenges facing the theory*. Phys.org. https://phys.org/news/2023-03-multiverse-tackling-theory.html

Manyika, J., Silberg, J., & Presten, B. (2019, October 25). *What do we do about the biases in AI?* Harvard Business Review. https://hbr.org/2019/10/what-do-we-do-about-the-biases-in-ai

McKinsey & Company. (2024). *The state of AI in early 2024: Gen AI adoption spikes and starts to generate value.* https://www.mckinsey.com/capabilities/quantum black/our-insights/the-state-of-ai

Moreno, R., & Park, B. (2012, June 5). *Cognitive load theory: Historical development and relation to other theories.* Cambridge University Press.

Nonlocality and entanglement. (n.d.). The Physics of the Universe. https://www.physicsoftheuniverse.com/topics_quantum_nonlocality.html

Paulson, L. C. (n.d.). *Isabelle: The next 700 theorem provers.* University of Cambridge. https://arxiv.org/pdf/cs/9301106

Pearce, T. (2023, May 4). Using generative AI to imitate human behavior. *Microsoft Research.* https://www.microsoft.com/en-us/research/blog/using-generative-ai-to-imitate-human-behavior/

Quantum mechanics and the puzzle of human consciousness. (2024, June 18). Allen Institute. https://alleninstitute.org/news/quantum-mechanics-and-the-puzzle-of-human-consciousness/

Rushkoff, D. (2010). *Program or be programmed: Ten commands for a digital age.* OR Books

Smith, C., McGuire, B., Huang, T., & Yang, G. (2006). *The history of artificial intelligence.* University of Washington. https://courses.cs.washington.edu/courses/csep590/06au/projects/history-ai.pdf

Solomonoff, G. (2023, May 6). *The meeting of the minds that launched AI.* IEEE Spectrum. https://spectrum.ieee.org/dartmouth-ai-workshop

Solomons, M. (2023, December 10). 70 UX statistics: Data analysis and market share. *Linearity Blog.* https://www.linearity.io/blog/ux-statistics/

Stevens, C. (2023, November 6). *The effect of AI on identity.* EU Reporter. https://www.eureporter.co/internet-2/artificial-intelligence/2023/11/06/the-effect-of-ai-on-identity/

The 10 biases you should know to break the bias in analytics. (n.d.). IAPA. https://www.iapa.org.au/news-and-articles/the-10-biases-you-should-know-to-break-the-bias-in-analytics.html

TSNE. (n.d.). Scikit-learn. https://scikit-learn.org/stable/modules/generated/sklearn.manifold.TSNE.html

UMAP: Uniform manifold approximation and projection for dimension reduction. (n.d.). UMAP Learn. https://umap-learn.readthedocs.io/en/latest/

Vancar, P. (2023, November 29). *Percentage of U.S. adults who agreed there is less stigma against people with mental illness than there was 10 years ago as of 2021.* Statista. https://www.statista.com/statistics/1104513/us-adult-opinion-mental-health-stigma/

Vena Solutions. (2024, July 18). *80 AI statistics shaping business in 2024.* https://www.venasolutions.com/blog/ai-statistics

What is strong AI? (2021, October 13). IBM. https://ibm.com/topics/strong-ai

Wingmore, I. (n.d.). *What is 70 percent rule for productivity?* TechTarget. https://www.techtarget.com/whatis/definition/70-percent-rule

Yasar, K., Chai, W., & Wigmore, I. (n.d.). *What is a MOOC (massive open online course)?* TechTarget WhatIs? https://www.techtarget.com/whatis/definition/massively-open-online-course-MOOC